NATIONAL DRAWING COURSE.

TEXT-BOOKS.

Free-Hand Drawing.
Mechanical Drawing.
Color Study.
Light and Shade.

TEACHERS' MANUALS.

Outline of Drawing Lessons for Primary Grades.
Outline of Drawing Lessons for Grammar Grades.

DRAWING CARDS.

National Drawing Cards for Primary Grades.

DRAWING BOOKS.

One book each for the 4th, 5th, 6th, 7th, and 8th
years of school.

SPECIAL MATERIAL FOR THE NATIONAL DRAWING COURSE.

The Cross Transparent Drawing Slate.
The Cross Pencil for use with the slate.
The National Drawing Models.
The National Model Support or Desk Easel.

National Drawing Books

MECHANICAL DRAWING

A MANUAL FOR

TEACHERS AND STUDENTS

BY

ANSON K. CROSS

Instructor in the Massachusetts Normal Art School, and in the School of Drawing and Painting, Museum of Fine Arts, Boston. Author of "Free-Hand Drawing, Light and Shade, and Free-Hand Perspective," and a Series of Text and Drawing Books for the Public Schools.

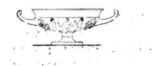

GINN & COMPANY

BOSTON · NEW YORK · CHICAGO · LONDON

The Athenæum Press

GINN & COMPANY PRO-
PRIETORS BOSTON U S A

PREFACE.

THE following notes are intended for students, and for teachers of elementary work, particularly for public school teachers There are many books on the subjects of projection and working drawings, but none which present the principles in ways suited to the needs of the large number of teachers who are required to give instruction in these subjects in the public schools Most of these teachers have had little instruction in the subjects and frequently do not understand problems which they are expected to explain This occurs because too difficult work is often planned for their grades, and also because the instruction which many public school teachers have received has been so advanced and theoretical that the simple principles which alone are necessary for elementary work have been lost in the attempt to understand descriptive geometry and the drawings of machine and other details, whose nature and use are often not known.

To understand descriptive geometry certain qualities of mind are absolutely necessary, and many find it impossible to comprehend even the simpler problems of this subject. The draughtsman or drawing teacher of advanced work who is without practical knowledge of the subject of descriptive geometry is very poorly equipped for his duties; but this knowledge is not necessary for the public school teacher, who will find it best to treat the subject of working drawings in a much simpler way.

This book presents principles and not a graded course of lessons. It covers more than many teachers may require, though special students or classes of the high school may study work as advanced as any of that presented. Teachers of drawing in the high school should understand all the problems of the book.

It is hoped that the book may assist teachers of both grammar and high schools so to understand the subject that they may give to classes instruction suited to their capacity and needs.

The author desires to express his obligations to G. A. HILL, A.M., who has read the proof-sheets and furnished him with many valuable suggestions.

ANSON K. CROSS.

CONTENTS.

————◦•◦————

MECHANICAL DRAWING.

CHAPTER I

MATERIALS AND THEIR USES.

1. Good work cannot be done without good instruments. The best work cannot be done without steel T-squares and triangles, steel-edged drawing boards, and drawing instruments of the best make. Students of art and technical schools should provide themselves with the best instruments.

In the public schools, no more can be done than to give a little knowledge of the principles of instrumental drawing, which will be valuable to all. To do the best work technically is impossible, because the pupils are too young, because they do not have the time necessary for practice, and because they do not have the materials necessary to produce the best work. Any one of these reasons is sufficient to prevent the making of first-class drawings, and together they make it impossible for us to expect that perfect results can be obtained.

Although pupils of the public schools labor under the above-mentioned disadvantages, it will not do for the teacher to expect bad work, or to be satisfied with work which at first glance is seen to be inaccurate; for even with the imperfect materials provided it is possible, by careful work, to obtain drawings which are neat and sufficiently accurate for the requirements of simple working drawings.

2. Paper. — The paper must be tough and should have a surface which is not easily changed or roughened by erasing lines drawn upon it. This is most important when drawings are to be inked. For free-hand sketching a soft paper is best, but for all mechanical work, the paper should be hard and strong.

For pencil drawings a paper which is not smoothly calendered is best, because the pencil marks more readily upon an unpolished paper, and because its surface will not show erasures as quickly as that of a smooth paper. For public school use, several kinds of cheap paper, which are good enough for the work, may be obtained both in sheets, in block form, and also made up in blank books

Whatman's paper is the best for drawings which are to be inked. There are two grades, hot and cold pressed, suitable for this use; the cold-pressed having the rougher surface   If the paper is not to be stretched, the cold-pressed is preferable, as its surface shows erasures less than that of the hot-pressed.   The side from which the water-marked name is read is the right side, but there is little difference between the two sides of hot and cold pressed papers   Stretching the paper is unnecessary except when colors are to be applied by the brush, or when the most perfect inked drawing is desired.

The use, in the public schools below the high, of sheets of blank paper instead of blank or other drawing books has its advantages and its disadvantages. If a drawing upon a sheet of paper is spoiled, it may be thrown away and another begun upon a new sheet; but this fact tends to careless work.   To the teacher loose sheets of paper are a source of great care, even if kept in portfolios or large envelopes and retained by the pupils, for the drawings must be examined, and it is not easy to keep them arranged in the order in which they are made

Blank books cost little more than paper, and their use tends to neatness and care on the part of the pupils, each of whom is interested to produce the best book.   Drawings in the books are always arranged in order and ready for examination.   The chief objection to their use is that they cannot be handled as a block for free-hand purposes, or be used with a T-square for instrumental work   This difficulty is avoided by fastening the book, by means of two rubber bands, to a drawing board made to receive it   When fastened to this board the book may be used for instrumental drawing as advantageously as paper upon a board.   Used in this way a book is preferable to sheets of paper; therefore a board should be provided whenever books are used, whether the books are blank books or those of any system

3. **Drawing Boards.** — Drawing boards should be made of clear white pine, and should not be painted or varnished; they should have cleats upon the back, so that upon the whole working surface

of the board the grain of the wood runs in one direction ; for when paper is stretched upon cleats which are fastened at the ends of the board, it is often spoiled by the swelling of the board, which moves back and forth upon the cleats.

The cleats should not be glued or otherwise firmly secured to the board, since the board must change its width with the weather, and if the cleats are firmly secured to it, the board will split or warp. The cleats should be fastened by means of screws and washers, the screws being placed in slots, which allow the board to move. This construction is not necessary for small boards required in the grammar schools.

The best board for use with a drawing book is one provided with a groove to receive the head of the T-square, which should be thick enough to remain in the groove when the book is between the square and the board.

4. Pencils. — To do good instrumental work two grades of pencils should be used, a hard one for the fine working lines and a softer one for the result lines. The hard pencil should be sharpened to the wedge-shaped point illustrated, which is as thin as possible one way, and about one-half the width of the lead the other.

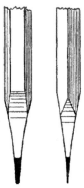

The softer pencil should have a point of the same shape but thicker, so as to give the width required for the result line.

To sharpen the pencil a knife should be used to give the wood the flattened form required in the lead. The lead should then be worked down by the use of a fine file. A substitute for the file, which may be used in the public schools, is given by gluing a straight-edged piece of sandpaper to a strip of wood. A point with which good work can be done cannot be obtained by the use of the knife alone.

A pencil with a round point should be used for all free-hand lettering and figuring, and for other work of a free-hand nature, such as the drawing of irregular curves.

The hard pencil should be used very lightly, as much pressure will indent the paper so that its marks cannot be removed.

All lines should be drawn with the pencil slightly inclined in the direction in which it is moved. The pencil should not be moved in the opposite direction, as it will then act as a plough to tear the surface of the paper.

Any and all lines not needed in the finished drawing should be erased at one time after the final lines have been determined, for the surface of the paper is soiled very quickly when worked upon, after erasures have been made. The working lines and other lines that are to be removed, should be erased when the drawing is ready to finish and before its outlines have been strengthened, in order that the final lines may be left in perfect condition, and may not require retouching on account of use of the eraser.

5. T-Square. — A T-square is an instrument used in connection with a drawing board, for drawing straight lines, it consists of two parts, a blade and a head, which are secured at right angles to each other. The blade of the T-square should be placed upon the head, which should never be cut to receive it, and should be secured to it by means of screws. This construction allows the parts to be separated for straightening, and the triangles to be moved across the head, which is often desirable, but cannot be done when the blade is set into the head.

The T-square should be used for drawing horizontal lines only. Its head should always be placed upon the left edge of the board. Vertical lines should be drawn by the use of a triangle placed upon the T-square and not by means of the T-square only; because the edges of the board are seldom exactly at right angles to each other, and the blade of the T-square is often not at right angles to the head, so that lines at right angles to each other will not result from the use of the T-square upon all the edges of the board. Only the upper edge of the T-square should be used, as the edges are often not quite straight or parallel.

6. Triangles. — The usual forms are illustrated The 45° triangle has two angles of 45° and one of 90° The 30° and 60° triangle has an angle of 30°, one of 60°, and one of 90°. By placing these triangles upon the T-square, lines at any of these angles with a vertical or horizontal line may be drawn.

Lines parallel to any given line, AB, may be drawn by first placing together two triangles (C and D), so that a side of one (C) coincides with AB, and then sliding triangle C upon triangle D, being
careful not to allow D to move.

Any parallel lines are most conveniently drawn by sliding one
triangle upon the other, or upon the T-square.

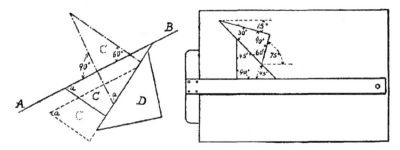

Lines perpendicular to AB may be drawn by placing either triangle upon the other, so that its hypotenuse coincides with AB,
and then revolving the triangle through an angle of 90°, into the
position illustrated by the dot and dash lines.

Lines at 30°, 45°, and 60° with AB, may be drawn by placing one
triangle upon the other, so that an edge coincides with AB, and then
reversing the triangle, as will be shown by experiments.

When paper is to be cut by the use of a knife and straight edge,
the T-square or triangle should not serve as the straight edge, for
its edges would soon be nicked and spoiled by the knife.

7. Compasses. — Compasses suitable for any use should have
jointed legs, which will allow the points to be placed at right angles
to the paper, whatever the size of the circle to be drawn. Compasses
should not be used for circles which are too large to allow the points
to be thus placed. A lengthening bar is generally provided, which
greatly increases the diameters of circles which may be drawn.

The. joint at the head of the compasses is the most important
feature. It should hold the legs firmly in any position, so that in
going over a circle several times, only one line will result. It should
allow the legs to move smoothly and evenly, and should be capable

of adjustment, as the parts soon wear, and the joint then becomes too loose. The joint should never be so tight that much pressure is required to move the legs.

One leg of compasses is usually provided with a socket to which are fitted three points ; a divider point, a pencil point, and a point carrying a special pen for the inking of circles. Each of these points is generally provided with a joint, so that it may be placed at right angles to the paper.

The other leg should be jointed ; it is often provided with a socket which receives two points, one a divider point, and the other carrying a needle point. Such an instrument may be used as dividers for spacing, or as compasses for penciling or inking circles, and will be all that is needed for public school work. The pen point is not necessary for grammar school use, as inking should not be attempted before the high school.

The cheaper grades of compasses are provided with sharp, pointed, solid legs, and are very objectionable, because these points work into the paper as the compasses revolve and make large holes which spoil the drawing. Even for public school work, compasses should have a needle point with a shoulder which prevents the making of holes in the paper. It may not be possible to provide for the grammar schools compasses with jointed legs ; but the needle point with the shoulder must be insisted upon, if even fair work is desired.

The compasses should be held very lightly between the thumb and forefinger, and should be inclined slightly in the direction in which the line is drawn. No more pressure should be applied than is necessary to obtain the line. In inking, little more than the weight of the compasses is needed ; but in penciling more pressure will be required for result lines Two grades of pencils are more necessary for use in the compasses than for straight line work ; they should be sharpened as shown on page 3, and so that the wide side of the point is at right angles to a line extending from the pencil point to the needle point. As this is difficult to do, pupils below the high school may use a sharp conical point.

The legs of the compasses should be moved, to give any desired radius, by taking hold of each leg. The needle point may be placed on one point of the drawing while this is done , but if the strain due to changing the radius is brought upon the needle point, it will tear the paper and spoil the drawing.

8. **Dividers.** — The compasses are changed into dividers as already explained. To set off equal spaces on any line, hold the dividers lightly between the thumb and forefinger and place one divider point at any desired point in the line to be spaced , then revolve the dividers about this point as centre until the other divider point comes to the line, when the dividers are to be revolved about it as about the first point. Revolve the divider points first on one side and then on the other of the line to be spaced, and never lift both points from the paper at the same time. Continue thus until the line is spaced.

To divide a line of given length into any number of equal parts, by use of the dividers, it will be necessary to use so little pressure that visible punctures are not made until the correct space has been obtained , if this is not done in two or three trials a second line should be drawn and the spacing continued upon it   If one line is gone over many times, the divider points will fall into the punctures previously made, and accurate work will not be obtained   When the correct space is found the points must be marked in the line so that they can be readily seen. To do this the line should be gone over several times, each time more pressure being applied to the dividers, until the punctures become visible. If the pressure required to place the points is applied at once, equal spaces will not be given, as the dividers will spring and move while marking the points

To do work of this nature easily, a pair of spring dividers should be used   This instrument has one point attached to a spring, which is regulated by a screw, so that very slight changes in the space may be made with ease.

9. **Needle.** — The needle may be used by the draughtsman and by advanced pupils with great advantage as regards both accuracy and speed. It should not, however, be used by young pupils, or below the high school.

The needle may be used to set off distances from the scale and to mark the intersections of lines, it may also be used when lines are to be drawn through two points, to hold the triangle so that only one point requires the attention of the draughtsman. This is done by placing the needle in one point, holding the triangle against it, and revolving the triangle until it comes to the second point.

Points should never be marked by holes in or through the paper, but by the smallest punctures which can be seen. These are much more definite than pencil marks, and have the advantage of locating points so that they are not lost by erasures. When a point is marked in this way a small pencil circle should be drawn about it free-hand, in order that its position may be readily seen.

The needle point may be placed in a handle of soft wood, and should project just far enough to be used, but no farther, as accidents will happen if it is not carefully handled. The best place for the needle point is at the unsharpened end of a pencil A double-ended pencil holder for round leads may have the pencil in one end and the needle in the other. This holder should be about $4\frac{1}{2}''$ long, so that it may be reversed readily.

10. **Scales.**— When objects are small they may be represented full size, but when large, the drawings must be smaller. Common scales for mechanical drawings are $\frac{1}{2}$, $\frac{1}{4}$, $\frac{1}{8}$, and $\frac{1}{16}$ full size These scales are often written $6''=1$ ft., $3''=1$ ft , $1\frac{1}{2}''=1$ ft. and $\frac{3}{4}''=1$ ft.

Large objects are often drawn to very small scales, in maps an inch often represents many miles.

Instead of selecting one of the scales named or one found upon the ordinary scales used by draughtsmen, drawings may be made to any scale whatever Thus, if any object is to be represented in a certain space, a scale should be constructed which will cause the drawing to fill the space in the best way.

To determine the scale by which the drawing of any object may be made of any desired size, divide the length of the object by the length of the drawing desired. Thus, suppose an object $21''$ long is to be represented in a space which will allow the drawing to be $10\frac{1}{2}''$ long. The drawing must be half size, and may be made by measuring the lines of the object and making those of the drawing half as long. If the drawing can be but $7''$ long, it will be one-third size. To make the scale for this drawing, draw a line 4″ long and divide it into twelve equal parts, which represent inches. Divide one of these spaces into eight equal parts to represent eighths of inches. By means of this scale, the drawing may be readily made by taking from it the dimensions of the different parts.

If views are to be made of a mallet, the length of the front view to be 6″, while the actual length of the object is 15″, a scale from which the sizes of the different parts can be taken may be made by drawing a line six inches long and dividing it into fifteen equal parts, each part representing one inch. One space may be divided into eight equal parts, and by means of this any part of an inch may be obtained.

The triangular scale (architects') has upon it the following scales : $\frac{3}{32}$, $\frac{1}{8}$, $\frac{3}{16}$, $\frac{1}{4}$, $\frac{3}{8}$, $\frac{1}{2}$, $\frac{3}{4}$, 1, 1$\frac{1}{2}$ and 3 inches. These are not provided and are not necessary in the grammar schools, where a cheap foot rule having three or four scales ranging from $\frac{1}{4}$ to $\frac{3}{4}$ full size will generally be all that is required. If a drawing cannot be made full or half size, or by one of these scales, a scale can be constructed by the pupils, as explained.

When views of objects are to be made, they should be of such size and so arranged as to produce a pleasing effect upon the sheet or the page, and when none of the scales provided will give the best size to the drawing, a special scale should be constructed.

11. Irregular or French Curves. — These curves are not required below the advanced work of the high school. They are used for curves which cannot be drawn with the compasses. To draw an ellipse or other curved line by their use, care must be taken to have the French curve exactly cover as many points in the line as possible, and then only the central points should be connected. If the line is drawn to all the points covered by the French curve, it will generally fail to continue smoothly through the points in the line beyond the curve.

When French curves are to be used the lines should be very lightly sketched free-hand through the determined points, before the curve is applied to the drawing to obtain the final lines.

12. Penciling. — Penciling is generally done to prepare for inking or for the making of tracings on tracing cloth, but sometimes practical shop drawings are finished in pencil lines. In this case the drawing is generally little more than a rough diagram, hastily made, and not intended for continued use.

Drawings finished in ink are much more effective and desirable than pencil drawings; but, as a good inked drawing cannot be made except upon an accurate pencil drawing, pupils should begin with the pencil, and should not be allowed to use ink until they can produce satisfactory results in pencil.

The aim in pencil work should be to approach as nearly as possible the accuracy and decision which are only attainable in an ink drawing. Care in selecting and using pencils, paper, and instruments will produce pencil drawings which have firm, even, black lines, that are very effective, if they are not quite so strong as those made with ink.

It is customary to represent all visible edges by full lines, and all invisible edges by dotted lines. These dotted lines should be as strong as the full lines; they should be regular and composed not of dots, but of very short dashes, whose lengths are uniform and greater than the spaces between them. Dotted lines are often very poorly drawn, and spoil the effect of drawings otherwise good. The width of the line and the length of the dash depend upon the size of the drawing and its purpose. For the drawings which pupils should make, the line given may serve as a copy.

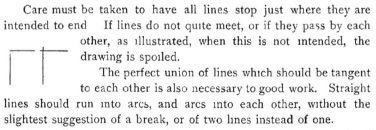

Care must be taken to have all lines stop just where they are intended to end. If lines do not quite meet, or if they pass by each other, as illustrated, when this is not intended, the drawing is spoiled.

The perfect union of lines which should be tangent to each other is also necessary to good work. Straight lines should run into arcs, and arcs into each other, without the slightest suggestion of a break, or of two lines instead of one.

Centre lines are necessary in working drawings. In the study of projection, it is not necessary to show them unless dimensions are to be given. They are represented by dot and dash lines, as illustrated, and extend some slight distance outside the drawing, to show

———.———.———.———.———.———.

that they are centre lines. Dimensions are placed as explained in Art. 44.

Different materials may be conventionally shown in pencil drawings by using different kinds of section lines. The materials most commonly used are often represented as illustrated in Art. 76.

An accurate and neat drawing is very pleasing and effective, every draughtsman and advanced student should be able to produce such a drawing with ease. Neatness and accuracy, the results of care and long practice, are essential to good work, and pupils should understand that without the exercise of much patience, work of real value cannot be obtained.

CHAPTER II.

GEOMETRICAL PROBLEMS.

THE following problems are those most likely to prove valuable to grammar and high school pupils. They are given in order that pupils may be able to apply them in the study of design and working drawings, and that they may learn to make accurate drawings. They are not given with the intention of teaching geometry.

The illustrations represent the working lines by light lines, the given lines by lines of medium strength, and the result lines by heavy lines. In finishing pencil drawings the best effects are obtained by using dotted lines for the working lines; but if time necessary to obtain perfectly regular lines cannot be given, it is better to use fine, light lines. When drawings are inked the working lines may be red, the given lines blue, and the result lines black.

In working all problems, to obtain accurate results, arcs described as working lines should be of large radii.

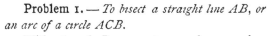

Problem 1. — *To bisect a straight line AB, or an arc of a circle ACB.*

With *A* and *B* as centres and any radius greater than half *AB*, describe arcs which intersect at *1* and *2*. Join *1* and *2* by a straight line, and *1-2* is perpendicular to *AB*, and bisects it in *3*, and the arc in *C*.

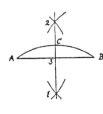

Problem 2. — *To erect a perpendicular to a given line at a given point A in the line.*

With *A* as centre and any radius, set off equal distances *A 1* and *A 2* from *A*. From points *1* and *2* as centres, and with a radius greater than half *1-2*, describe arcs which intersect in *3* and *4*. Join *3* and *4*, and *3-4* is the required perpendicular.

Problem 3. — *To draw a perpendicular to a given line BC, from a point A outside the line.*

With *A* as centre and any radius, intersect the given line in points *1* and *2*. With points *1* and *2* as centres and any radius describe arcs intersecting in *3*. Join *A* and *3*; *A 3* is the required perpendicular.

Problem 4. — *To erect a perpendicular to a given line AB, from a point B, at or near its end.*

With *B* as centre and any radius, draw an arc of a circle *1–2–3*. With *1* as centre and the same radius, cut this arc in *2*, and with *2* as centre and the same radius, describe the arc *3–4*. With *3* as centre and the same radius, intersect *3–4* in *4*. Join *4 B*; this is the required perpendicular.

Problem 5. — *To draw a line, EF, parallel to a given line, AB, and the distance CD from it.*

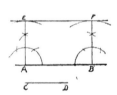

From any two points, *A* and *B*, in the line as centres, with radius *CD*, describe arcs *E* and *F*. Erect perpendiculars at *A* and *B*, which intersect the arcs in *E* and *F*. Join *E* and *F*; this is the required parallel. In practice, it is not necessary to draw the perpendiculars.

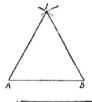

Problem 6. — *To construct an equilateral triangle on a given base, A B.*

With *A* and *B* as centres and *A B* as radius, describe arcs which intersect at *1*. Join *1 A* and *1 B.*

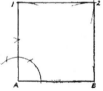

Problem 7. — *To construct a square on a given base, AB.*

Draw *A 1* perpendicular to *AB* and equal to it. (Problem 4.) With *B* and *1* as centres and radius *AB*, describe arcs intersecting in *2*. Join *1–2* and *2 B.*

Problem 8. — *To construct a rectangle of given sides, AB and CD.*

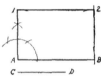

At *A*, by Problem 4, erect a perpendicular *A1* equal to *CD*. With *1* as centre and radius *AB*, describe an arc, and intersect this arc in *2*, by one described from *B* with *CD* as radius. Join *1–2* and *2 B*.

Problem 9. — *To inscribe a square within a given circle.*

Draw *AB*, a diameter of the circle, which will be a diagonal of the square. By Problem 1 bisect *AB*, and continue the bisector to inter-sect the circle in *1* and *2*. Join *A 1, 1 B, B 2,* and *2 A.*

NOTE. — A square may be constructed upon a given diagonal by draw-ing a circle through its ends, *A* and *B*, and proceeding as explained above.

Problem 10. — *To bisect a given angle, CAB.*

With *A* as centre and any radius, describe an arc intersecting *AC* and *AB* in points *1* and *2*. With points *1* and *2* as centres and any radius, describe arcs which intersect in *3*. Join *A* and *3*.

Problem 11. — *To trisect a right angle, CAB.*

With *A* as centre and any radius, describe an arc intersecting *AC* in *1* and *AB* in *2*. With *1* and *2* as centres and the same radius intersect the arc in *3* and *4*. Join *A 4* and *A 3*.

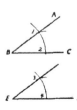

Problem 12. — *To construct at E, in the line E D, an angle equal to a given angle, ABC.*

With *B* as centre and any radius, intersect *AB* and *BC* in *1* and *2*. With *E* as centre and same radius describe an arc intersecting *E D* in *4*. With *4* as centre and *1–2* as radius, inter-sect the arc in *3*. Join *E* and *3*.

Problem 13. — *To construct an angle of 90°, 60°, 45°, 30°, 15°, or any other given magnitude.*

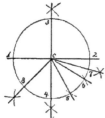

There are 360° in the entire circumference of a circle. Draw a diameter *1–2*, and there are 180° on either side of this line. Draw a second diameter *3–4*, at right angles to *1–2*, and the circle is divided into four equal angles of 90°. Trisect one of these by Problem 11, and angles of 30°, *2 c 6*, and of 60°, *2 c 5* are obtained. Bisect one of these angles of 30°, and angles of 15°, *2 c 7*, and *7 c 6* are obtained. Bisect an angle of 90° and angles of 45° are obtained. Trisect, by spacing, an angle of 15° and angles of 5° are given. Divide one of these into five equal parts and degrees are given. In this way any angle may be obtained.

Problem 14. — *To divide a given line, A B, into any number of equal parts, as five.*

Draw *AC* at any angle to *AB*. On *AC* set off any distance five times. Join *B* and the 5th point, and through the other points draw parallels to *B 5* (Art. 6), which will divide *AB* as required.

Problem 15. — *To construct a triangle, having given its sides, AB, C, and D.*

With *A* as centre and radius *C*, describe an arc. With *B* as centre and radius *D*, intersect this arc in *1*. Join *1 A* and *1 B*.

Problem 16. — *To inscribe a regular hexagon within a given circle.*

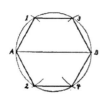

Draw the diameter *AB* of the circle, which will be a long diagonal of the hexagon. With point *A* as centre and the radius of the circle as radius, intersect the circle by arcs *1* and *2*. With *B* as centre and same radius, intersect the circle by arcs *3* and *4*. Join *A–1*, *1–3*, *3–B*, *B–4*, *4–2*, and *2–A*.

Note.—By joining every other point an equilateral triangle will be obtained.

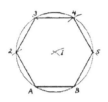

Problem 17. — *To construct a regular hexagon upon a given base AB.*

With *A* and *B* as centres and *AB* as radius describe arcs which intersect at *1*. With *1* as centre and same radius, describe a circle, and in its circumference place points *2, 3, 4,* and *5,* which are equidistant and the distance *AB* apart. Join *A–2, 2–3, 3–4, 4–5,* and *5 B.*

Note. — The radius of any circle applied, as a chord, six times to the circumference divides it into six equal parts.

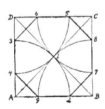

Problem 18. — *To construct a regular octagon within a given square ABCD.*

Draw the diagonals of the square, which intersect at its centre, *1,* and with points *A, B, C,* and *D* as centres and a radius, *A 1,* equal to half the diagonal, describe arcs intersecting the sides of the square in points *2, 3, 8, 9, 6, 7, 4,* and *5.* Join *4–9, 2–7, 8–5,* and *6–3.*

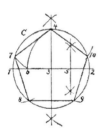

Problem 19. — *To inscribe a regular pentagon within a given circle C.*

Draw the diameter *1–2* and a radius *3–4* perpendicular to it. Bisect *2–3* in *5,* and with *5* as centre and radius *5–4,* intersect *2–1* in *6.* With *4* as centre and radius *4–6* intersect the circle in *7.* Set off the distance *4–7* from *7* to *8, 8* to *9, 9* to *10.* Join *4–7, 7–8, 8–9, 9–10,* and *10–4.*

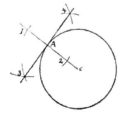

Problem 20. — *To draw a tangent at any point A, in a given circumference.*

Draw a radius, *cA,* and erect a perpendicular, *3–4,* to the radius at *A,* by Problem 2. This is the required tangent.

Problem 21. — *To inscribe a circle within a given triangle ABC.*

By Problem 10, bisect any two angles of the triangle, as *CAB* and *ABC*. The bisecting lines intersect at *1*, the centre of the required circle. A perpendicular, as *1–2*, from *1* to any side of the triangle, is the radius of the required circle.

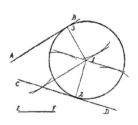

Problem 22. — *To draw an arc of a given radius EF, tangent to given straight lines AB and CD.*

Draw parallels to *AB* and *CD* by Problem 5, at the distance *EF* from them. The parallels intersect at *1*, the centre of the required arc. The arc is tangent to *AB* and *CD*, at points *3* and *2*, in perpendiculars to these lines from *1*.

Problem 23. — *To draw an arc of a given radius DE, tangent to a straight line AB and a circle C.*

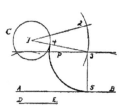

Draw by Problem 5 a parallel *P 3* to *AB*, the distance *DE* from it. With *1*, the centre of circle *C*, as centre and a radius *1–2*, equal to the radius of the given circle plus *DE*, that of the required arc, describe an arc to intersect the parallel *P 3* in *3*, the centre of the required arc. A perpendicular from *3* to *AB* gives *5*, the point of contact of the arc and *AB*, and a straight line from *3* to *1* gives *4*, the point of contact of the arc and the circle.

Problem 24. — *To draw an arc of a given radius CD tangent to two circles A and B.*

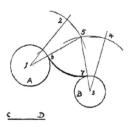

With *1*, the centre of the circle *A*, as centre and radius *1–2*, equal to the radius of *A* plus *CD*, describe an arc *2–5*. Intersect this arc in *5* by one described from *3*, the centre of the circle *B*, with a radius *3–4*, equal to the radius of *B* plus *CD*. Point *5*

is the centre of the required arc, and lines from *5* to *1* and from *5* to *3* give *6* and *7*, the points of contact of the arc and circles *A* and *B*.

Problem 25. — *To draw, within a given equilateral triangle ABC, three equal circles, each tangent to two others and to one side of the triangle.*

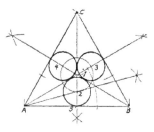

Bisect the angles of the triangle and obtain *1*. Bisect the angle *1 AB* and obtain *2*, the centre of one circle. With *1* as centre and radius *1–2* draw an arc to obtain centres *3* and *4*. With centres *2*, *3*, and *4* and radius *2–5*, describe the required circles.

Problem 26. — *To draw arcs of circles tangent at points C and B, to two parallel straight lines AB and CD and passing through any point as E, in the line CB.*

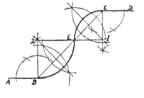

At *C* and *B* erect perpendiculars. Bisect *BE* and *CE*, continuing the lines of bisection until they meet the perpendiculars from *B* and *C* in points *2* and *1*, which are the centres of the required arcs. With *1* as centre and *1 C* as radius draw the arc *CE*, and with *2* as centre and *2 B* as radius draw the arc *BE*.

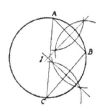

Problem 27. — *To circumscribe a circle about a given triangle ABC.*

By Problem 1 bisect any two sides as *AB* and *BC* by perpendiculars meeting at *1*. With *1* as centre and radius *1 A* describe the circle.

NOTE 1. — If any three points as *A*, *B*, *C* not in the same straight line are given, a circle may be passed through them by connecting the points and proceeding as above.

NOTE 2. — The centre of any circle may be found by assuming any three points in it and proceeding as above.

Problem 28. — *To construct a regular octagon on a given side AB.*

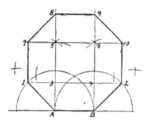

At *A* and *B* erect perpendiculars by Problem 4. Continue *AB* and bisect the outer right angles, making the bisecting lines *A 1* and *B 2* equal to *AB*. Draw *1–2* cutting the perpendiculars in *3* and *4*.

From *3* and *4* set off the distance *3–4*, giving *5* and *6*. Draw through *5* and *6* a straight line, and set off from *5* and *6* distances *5–7* and *5–8*, *6–9* and *6–10* equal to *3–1* or *3 A*. Join the points *1–7*, *7–8*, *8–9*, *9–10*, and *10–2*.

Problem 29. — *To inscribe, within a given equilateral triangle ABC, three equal circles, each tangent to two others and to two sides of the triangle.*

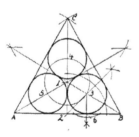

Bisect the angles *A*, *B*, and *C* by lines meeting at *1* ; bisect the right angle *C 2 B* to obtain *3*, the centre of one of the required circles. With *1* as centre and *1–3* as radius, describe a circle giving points *4* and *5*. *3*, *4*, and *5* are the centres of the required circles. The shortest distance, as *3–6*, from a centre to a side of the triangle, is the radius of the required circles.

Problem 30. — *Within a given square, ABCD, to draw four equal circles, each touching two others and two sides of the square.*

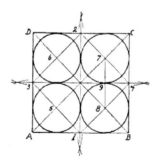

Draw the diagonals of the square to obtain its centre, and then lines parallel to the sides through the centre to intersect the sides in points *1*, *2*, *3*, and *4*. Join *1–3*, *3–2*, *2–4*, and *4–1*, obtaining points *5*, *6*, *7*, and *8*, the centres of the required circles. The length, *7–9*, of a radius is given by drawing *7–8*.

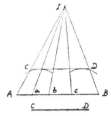

Problem 31. — *To divide a straight line, CD, into the same proportional parts as a given divided line, AB.*

Draw *CD* parallel to *AB* by Problem 5. Draw lines through *AC* and *BD* to meet in *1*. From points *a*, *b*, and *c* in *AB*, draw to *1* lines, which will divide *CD* as required.

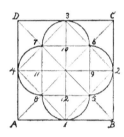

Problem 32. — *Within a square, ABCD, to draw four equal semi-circles, each touching one side, and forming by their diameters a square.*

Draw the diagonals and the diameters of the square and join points *1-2*, *2-3*, *3-4*, and *4-1*, obtaining *5*, *6*, *7*, and *8*. Join *5-6*, *6-7*, *7-8*, and *8-5*, obtaining *9*, *10*, *11*, and *12*, which are the centres of the required arcs, of which *9-2* is the length of a radius.

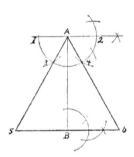

Problem 33. — *To construct an equilateral triangle, its altitude, AB, being given.*

At *A* and *B* erect perpendiculars to *AB* by Problem 4. With *A* as centre and any radius describe a semi-circle, which intersects the perpendicular in *1* and *2*, and from *1* and *2* set off the radius of the semi-circle in *3* and *4*. Draw *A 3* and *A 4* to intersect, in *5* and *6*, the perpendicular erected at *B*.

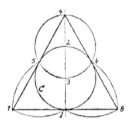

Problem 34. — *To circumscribe an equilateral triangle about a given circle C.*

Draw a diameter, *1-2*, of the circle. With *2* as centre and *2-3*, the radius of *C*, as radius, describe a circle intersecting *1-2* produced in *4* and the circle in *5* and *6*. With *1*, *5*, and *6* as centres, and radius *1-5*, describe arcs which intersect at *7* and *8*. Join *4 7*, *7-8*, and *8-4*.

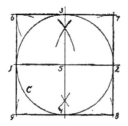

Problem 35. — *To circumscribe a square about a given circle.*

Draw the diameter *1–2*, and a diameter *3–4*, at right angles to *1–2*. With *1, 2, 3,* and *4* as centres and radius *1–5*, describe arcs which intersect at *6, 7, 8,* and *9*. Join *9–6, 6–7, 7–8,* and *8–9.*

Problem 36. — *To circumscribe a regular hexagon about a given circle C.*

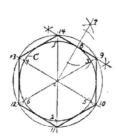

Draw a diameter, *1–2*, and with the radius of the circle set off, from *1* and *2*, points *3, 4, 5,* and *6* ; through these points draw radii, extending them beyond the circle. Bisect the sector *1 c 3* by *c 7*, which intersects the circle at *8*. At *8*, by Problem 2, erect a perpendicular and obtain *9* and *14*. With *c* as centre and *c 9* as radius, describe a circle to intersect the radii extended in *10. 11. 12,* and *13*, the vertices of the required hexagon. Join *9–10, 10–11. 11–12, 12–13,* and *13–14.*

Problem 37. — *To draw circles tangent to each other and to two lines, AC and BC, not parallel, the radius, DE, of one circle being given.*

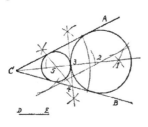

Bisect the angle *ACB* by *C 1*. Draw parallel to *AC*, and distant from it by the radius of the given circle, a line which, intersecting *C 1* at *2*, gives the centre of the first circle, which intersects *C 1* in *3.* At *3* erect a perpendicular to *C 1* and obtain *4*, and bisect the angle *3–4–C*, obtaining *5*, the centre of the second circle, whose radius is the distance *5–3*. In this way any number of circles may be drawn.

Problem 38. — *To construct a regular polygon of any number of sides (in this case seven), upon a given base, AB.*

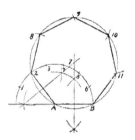

Extend *BA*, and with *A* as centre and *AB* as radius describe a semi-circle. Divide the semi-circle into as many equal parts as there are sides in the required polygon, and draw from *A* to the second point, a second side of the polygon. Bisect *A 2* and *AB*; the bisectors meet at *7*, the centre of the required polygon. With *7* as centre and *7 A* as radius, describe a circle and upon it set off the distance *AB*, beginning at *2*, and obtain points *8*, *9*, *10*, and *11*, the vertices of the required polygon. Join *2–8*, *8–9*, *9–10*, *10–11*, and *11 B*.

Problem 39. — *To inscribe a regular polygon of any number of sides (in this case five) within a given circle.*

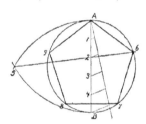

Draw a diameter *AB* and divide it into as many equal parts as the polygon is to have sides. With *A* and *B* as centres and radius *AB*, describe arcs intersecting in *5*. From *5* draw a straight line through the second point, *2*, to intersect the circle in *6*. *A 6* is a side of the required polygon. Beginning at *6*, set off *A 6* upon the circle to obtain points *7*, *8*, *9*. Join *A 6*, *6–7* *7–8*, *8–9*, and *9 A*.

NOTE. — This method is approximate only.

Problem 40. — *Within a given circle to draw any number of equal semi-circles tangent to the circle and forming by their diameters a regular polygon.*

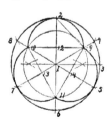

Draw radii *1–2* and *1–3* at right angles to each other. Beginning at *2*, divide the circle into twice as many equal parts as the number of semi-circles required, and draw diameters to these points. Join *2* and *3*; the intersection *9* of *2–3* with the diameter *4–7* is a vertex of the polygon whose sides are the diameters of the required semi-circles. Describe a circle through

9 with *1* as centre, and join its points of intersection, *9*, *10*, and *11*, obtaining the diameters which contain points *12*, *13*, and *14*, the centres of the required semi-circles, of which *12–2* is the length of a radius.

Problem 41. — *Within a given circle A to draw any number of equal circles tangent to each other and to the circle A.*

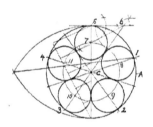

Divide the circumference of the circle into as many equal parts as inscribed circles are required, and draw radii to these points of division. Bisect the angle *1 c 5*, and at *5* draw a tangent which intersects the bisector at *6*. Bisect the angle *5–6–c* and obtain *7*, the centre of one of the required circles. With *c* as centre and *c 7* as radius, describe a circle to cut the radii previously drawn, in points *8*, *9*, *10*, and *11*, which are centres of the required circles. The radius of each circle is the distance *7–5*.

Problem 42. — *About a given circle A, to draw any number of equal circles tangent to each other and to the given circle.*

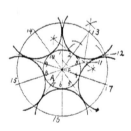

Divide the circumference of the circle into twice as many equal parts as the number of circles required, and draw radii extending them beyond the points of division. At any point, as *2*, draw a tangent to the circle intersecting a radius, extended, at *11*. Bisect the angle *2–11–12*, and obtain *13*, the centre of one of the required circles. With *c* as centre and *c–13* as radius, describe a circle to intersect every other radius in *14*, *15*, *16*, and *17*. These points are the centres of the required circles of which the radius is the distance *13–2*.

Problem 43. — *To draw an ellipse, its axes AB and CD being given.*

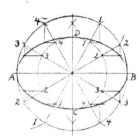

Upon the axes draw circles and divide both circles by diameters $1-1$; $2-2$, etc., into any number of parts, equal or unequal. From the points in the large circle draw parallels to the short axis, and from those in the small circle draw parallels to the long axis. The intersections of lines from points of the same number are points in the ellipse.

Problem 44. — *To draw, by means of a trammel, an ellipse, whose axes are given.*

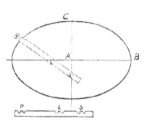

Set off AB, half the long axis, on the edge of a straight piece of paper from P to S, and AC, half the short axis, from P to L. Move this paper trammel so that S is upon the short axis and L upon the long axis, and P will always be a point in the ellipse. Its position may be marked by a sharp lead pencil. The best way is to use a needle point, and make very fine punctures in a straight line drawn near the edge of the trammel. The needle may then be placed through L and on the long axis, and the paper revolved about L till the point S is over the short axis, when the needle may be used to mark, through P, a fine puncture in the paper.

To enable the axes to be seen, the edges of the paper may be notched as indicated.

Problem 45. — *To draw upon given axes, AB and CD, an approximate ellipse.*

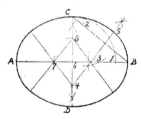

With the centre, 6, of the ellipse as centre and half the short axis, $6\ C$, as radius describe an arc $C\ 1$. Draw CB and from C set off $C\ 2$ equal to $B\ 1$, or the difference between half the short and half the long axis. Bisect $B\ 2$ and continue the bisecting line to intersect AB in 3,

and *CD* in *4*. With *3* as centre and radius *3 B* describe an arc
B 5, and with *4* as centre and *4–5* as radius, describe an arc which
will pass through *C* and complete one quarter of the curve. From
6, set off *6–7* equal to *6–3*, and *6–8* equal to *6–4*, and draw lines
joining *4–7*, *8–7*, and *8–3*, and corresponding to *4–5*. With
points *3*, *4*, *7*, and *8*, as centres, complete the ellipse as explained above.

Problem 46. — *To draw an equable spiral.*

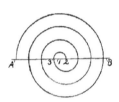

Draw *AB* and upon it place any two points,
1 and *2*, as centres. With *1* as centre, and
radius *1–2*, draw a semi-circle. With *2* as
centre and radius *2–3*, draw a semi-circle;
continue the process, using *1* and *2* as centres,
and the distance to the end of the diameter of
the semi-circle last drawn as radius.

Problem 47. — *To draw a variable spiral, its greatest diameter,
AB, being given.*

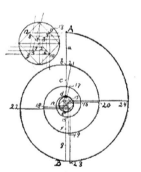

Divide *AB* into eight equal parts by
points *a*, *b*, *c*, etc. On *d e*, the fifth space,
as diameter, draw a circle which is called
the *eye* of the spiral. Inscribe a square in
the circle and draw its diameters, as shown
in the enlarged drawing of the eye. Divide
these diameters into six equal parts and
number the points as shown. With *1* as
centre and *1–13* as radius, draw the arc
13–14. With *2* as centre and *2–14* as
radius, draw the arc *14–15*. With *3* as
centre and *3–15* as radius, draw the arc
15–16; and so proceed, using all the points up to and including *12*,
as centres, and limiting each arc by a line drawn from its centre
through the centre of the next arc.

Problem 48. — *To draw the involute of a circle.*

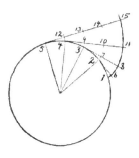

From any point, as *1*, in the circumference set off any number of equidistant points, *2, 3, 4,* and *5*. Draw tangents to the circle at these points, and on the tangent at *2* set off the distance *2–1*, giving *6*. On the tangent at *3*, set off the length of the arc *3–1*, giving *8*. At *4* set off the length of the arc *4–1*, giving *11*. At *5* set off the length of the arc *5–1*, giving *15*. A curve through *1, 6, 8, 11,* and *15* is the involute of the circle.

Note. — The distance between points 1, 2, 3, etc. must be such that the difference in length between the chord and the arc is slight.

Problem 49. — *To draw the cycloid curve traced by a point in the circumference of a given circle, A, which rolls on the line CD.*

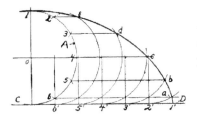

Divide the circumference into any number of equal parts, as twelve, and set off on *CD* the equidistant points in which the points marked in the circle will come to the straight line as the circle rolls upon *CD*. When the circle is tangent at *2'* its centre will be in a perpendicular erected at *2'* and in a parallel to *CD* through *o*, the centre of *A*. In the same way its centre will be found when it is tangent at 3', 4', 5', etc. As the circle rolls, point *1* describes the cycloid. In any position of the circle point *1* will be found by setting off as many equal spaces on the circle from points *2', 3', 4'*, etc., as the circle has rolled over, starting from *1'*. Thus, one space from *2'* to *a*; two spaces from *3'* to *b*, and so on, for the complete curve. Points *a, b, c*, etc., may also be found by noting the level of the points *6, 5, 4*, etc., in the given circle *A*. Thus, a horizontal line from *6* will give *a*; one from *5* will give *b*, and so on. In either of these ways, the complete curve, of which one-half only is shown, may be obtained. The length of the arc between the points in the circle of this and Prob. 48 may be computed, or a result practically correct obtained by dividing the circle into many parts.

Problem 50. —— *To construct the epicycloid curve traced by a point in the circumference of a circle, A, which rolls on the outside of a circle B.*

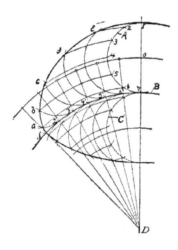

Divide the circumference of the circle *A* into any number of equal parts, as twelve, and set off from *7* on *B* the points *6', 5', 4', 3', 2', 1'*, in which the points *6, 5, 4, 3, 2, 1* of *A* will coincide with *B* as *A* rolls. Describe an arc of *A* when it is tangent, at each of the points *6', 5', 4', 3', 2'*; the centres will be in an arc concentric with *B*, passing through *o*, and in radii of B, extended through the points *6', 5', 4', 3',* and *2'*. When tangent at *2'* the marking point *1* is at *a*, and is obtained by setting off from *2'* one of the equal spaces into which it was divided, or by drawing an arc through *6*, with *D* as centre, to intersect at *a* the circle *A*, when tangent at *2'*. When tangent at *3'* the marking point *1* will be at *b*, and will be obtained by setting off two equal spaces from *3'* to *b*, or by an arc through *5*. In this way all the points, *c, d, e*, of the curve may be found ; also the other half of the curve.

Problem 51. —— *To construct the hypocycloid traced by a point in circle C, which rolls inside the circle B.*

The process is the same as for the outer epicycloid explained in Problem 50.

NOTE. —— When the diameter of the rolling circle is equal to the radius of the circle within which it rolls, the hypocycloid becomes a straight line and is a diameter of *B*.

CHAPTER III.

WORKING DRAWINGS.

13. Nature and use of Working Drawings. — The workman who builds a house, engine, or any other object, obtains the necessary information concerning the size, shape, kind of material, and amount of finish, etc, of all the different parts, from different projections or views of the object to be made. These views are called working drawings

14. Free-hand drawings, that is, perspective views, give the appearance of objects, but they represent only part of the surface of an object, and do not give the actual sizes or shapes of any of its parts ; therefore such drawings are not suitable for working drawings.

A perspective drawing may be used as a working drawing when the object illustrated is very simple, or of a form commonly used, and the dimensions of the parts are placed upon the drawing but when such a drawing is used the knowledge and experience of the workman supply the information which the perspective drawing does not give.

Generally, perspective views of machinery and architecture are made simply as illustrations of objects which either have not been made, or which are not placed so they can be photographed.   The best and cheapest perspective possible to obtain is given by a photograph

15. A working drawing must give the actual shapes of the parts it represents.   It may be the full size of the object or smaller, but it must be drawn to some determined scale and show the true proportions and relations of all the parts.   In projection the actual appearance of any object is given upon a plane at right angles to the direction in which the object is seen.   It follows that, as far as possible, working drawings should be made upon planes which are parallel to the principal faces of the object to be represented.

16. The nature and relations of the different views usually made as working drawings will be shown by the following experiments :

Place a cube, shown by Fig. 1, at the level of the eye so that only one face is visible, the corners of this face being equally distant from the eye. Hold a glass slate vertically in front of the cube and

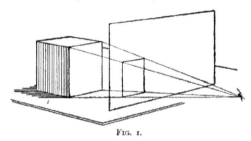

FIG. 1.

parallel to the visible face, that is, at right angles to the direction in which the cube is seen. Trace the appearance of the cube upon the slate. The tracing will be a square, for the apparent shape of any surface at right angles to the direction in which it is seen agrees with its real shape.

If the slate is now held horizontally above the top of the cube and the eye is placed above the object, so that the top face appears its real shape, a tracing of the cube from this position will be a square, as in the first case. The first tracing gives the real shape of the front face of the cube, and the second tracing the real shape of the top face of the cube. The tracing on the slate when vertical shows that the height of the object is equal to its width, while that upon the slate when horizontal shows that the distance from front to back is equal to the width. The two tracings together express the fact that the three dimensions of the cube are equal. This will be shown whatever the distances of the slate and the eye from the cube ; but to have the two tracings of the same size, the eye and slate must have the same relative positions with reference to the cube when making both.

17. The tracings are smaller than the cube because the visual rays converge toward the eye, and the slate is nearer the eye than is the cube. If the eye is placed farther away from the object, and the position of the slate remains the same, the tracing will be larger. If, as the tracing is made, the eye is moved so as to see every visible point of the object by means of a visual ray perpendicular to the slate, the tracing will be the full size of the object.

18. If instead of one slate which is held in the hand to receive the different views of the cube, two slates are used which are supported at right angles to each other, as in Fig. 2, and if instead of tracing the appearance when the eye is in one position, the eye is moved with the pencil point so that every point of the object is seen in a perpendicular to the slate, the tracing upon the vertical slate will give the real dimensions of the front face of the cube, and that upon the horizontal slate will give the real dimensions of the top face of the cube The two tracings will give all the actual dimensions of the object, and the dimensions of one tracing may be compared with those of the other. These tracings illustrate the nature of constructive drawings, and the slates in their different positions illustrate the imaginary planes upon which these drawings are supposed to be made

19. These tracings or drawings are called " projections" or " views," and when dimensioned they are called working drawings. The tracing upon the vertical slate is called the vertical projection, or the front view; that upon the horizontal slate is called the horizontal projection, or the top view. Working drawings are generally called views, which will be the term used throughout this book.

The pupil who has used the transparent drawing slate for freehand purposes will have no difficulty in making and understanding these experiments, he will readily see that the tracings made upon the vertical and horizontal slates give the three actual dimensions of the cube seen through them.

20. These drawings might be used to form the cube represented, by so shaping any material that, when it is looked at horizontally, it will cover the tracing upon the vertical slate, and when it is looked at from above, it will cover that upon the horizontal slate.

Objects are sometimes shaped by cutting them until they exactly cover the drawing which represents them this is the case particularly with irregularly curved parts. Such parts are, however, generally obtained by the use of templets, which are forms of thin wood or metal made from the drawings, and applied to the object to show when the desired form has been obtained In this way the frames of boats are shaped. also other irregularly curved objects

21. Objects are seldom shaped by placing them upon the drawings, since the drawings are often less than the full size of the object. Even when this is not the case it is difficult to obtain accurate work in this way.

22. Drawings which are intended for shop use always give in figures the dimensions of each part, also directions as to the material of which it is to be made, and the amount of finish which it is to receive. The machinist readily works to hundredths of inches; accuracy such as this, or even far less, would be impossible by such a clumsy method as placing the object upon the drawing.

23. The dimensioning of the drawings is to the workman of first importance, as he is not allowed to apply a scale to obtain a dimension omitted from the drawing, but to the student it is not at first important. His attention should be given to the study of the principles necessary to enable him to make drawings. When able to do this he should study dimensioning, which is advisable in the seventh and eighth grades.

24. Drawings are made upon paper, which cannot be used as was the glass slate to trace the appearance of an object, but as practical working drawings are not made by tracing, and since the chief use of a working drawing is to represent what exists only in the brain of the engineer, architect, or designer, working drawings would be of little service if they could be obtained only by tracing or by drawing from objects actually existing

Drawings represent so exactly the conceptions of the designer that the greatest mechanical constructions are made in small sections in the workshop, and are then taken away and put together, every screw and bolt hole of every part being properly placed, so that the "Ferris Wheel," or the engine of the steamship "Paris," or even the entire boat with all her fittings, can be put together the first time as surely as can the parts of a simple box.

The following paragraph from the *Century* for July, 1894, gives an idea of the importance of the draughtsman's duties.

"For the hull alone of the battleship Indiana, 25 principal plans must be made, and fully 400 separate drawings must be prepared, and duplicated by photographs This of itself is enough work to keep a force of expert draftsmen busy continuously for eight months. For the engines more than

250 separate drawings are required, and these in all their intricate details would take a force of 50 men nearly a year to complete, if engaged continuously at the task. Not only must every rivet and every joint be marked out and noted, but there must be the most complicated computation of strains and weights."

25. Study of Principles. — In order to understand readily the principles of the subject of working drawings, two slates should be hinged together so that they can be placed at right angles to each other, and can be revolved into the same plane, as shown by Figs. 2, 3, and 4.

Fig. 2 represents a cube and the vertical and horizontal glass slates, which represent the imaginary planes upon which the front

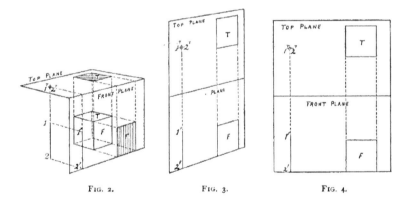

FIG. 2.   FIG. 3.   FIG. 4.

and top views are supposed to be made ; Fig. 3 represents the slates after the horizontal one has been revolved to coincide with the plane of the vertical slate ; Fig. 4 is a drawing which gives the real shapes of the two slates and of the views upon them.

26. It will be seen that after the slates have been placed as illustrated by Figs. 3 and 4, the views will be in line with each other, the top view being above the front view, and the two views of any point being in the same vertical line.

The drawings also show a vertical line **1–2,** and its views upon the two slates.

27. It often happens that these two views are not enough to describe an object. Thus, if the cube is bisected to form two triangular blocks, the views of either triangular block upon the vertical and horizontal slates will be the same as the views of the entire cube.

Fig. 5 represents a triangular block formed by bisecting the cube, illustrated in Fig. 2, by a plane passing through the front upper and

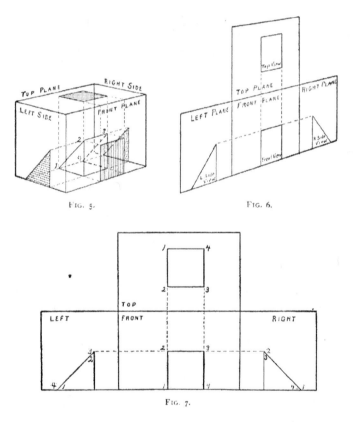

FIG. 5.   FIG. 6.

FIG. 7.

back lower edges; in Fig. 5 these edges are *2–3* and *1–4* respectively. It also represents side slates attached to the vertical and horizontal slates, and at right angles to both. If a tracing of the block is made upon one or both of these side planes, or slates, as upon the front and top slates, the information not given by the first two views will be supplied

Figs 6 and 7 represent the slates when the side slates and the top slate have been revolved to coincide with the plane of the front slate.

28. The views upon the side slates are at the same level as that upon the front slate, and when the slates are as shown in Figs 6 and 7, the right and left side views of any point are in a horizontal line through the front view of the point.

29. A fifth slate may be placed at the back, and thus a view of the back may be obtained. The surface upon which the object rests may be that of a sixth slate, and thus a view of the bottom may be traced. All these, together with additional views, are sometimes necessary This arrangement of slates and object really amounts to placing about the object a glass box and tracing upon the sides of the box, by means of perpendiculars to the sides, the different appearances of the object.

30. These perpendiculars, or visual rays, which produce the different views are called *projecting lines*. When the planes are revolved to coincide with the plane of the front plane, as shown by Figs. 6 and 7, these projecting lines are represented by the vertical and horizontal lines which contain the different views of the various points, and are the projecting, or working, lines of the drawings.

31. *The figures show that the different views are arranged with reference to the front view, so that the line of each view nearest the front view represents the front face of the object.*

By means of such glass slates hinged together, these simple principles may be so illustrated that pupils in the upper grammar grades can understand the nature of the drawings made for the different views of the various objects chosen for study, the reason for arranging working drawings as explained, also the fact that the front and top views of the same point are in the same vertical line, and the front and side views in the same horizontal line.

32. Some teachers prefer to think of the object as revolved so that its different surfaces are, one after another, placed parallel to one plane (or glass slate) upon which the drawings are made in the same manner as a drawing made upon any one of the slates at right angles to each other. There is no objection to this method, indeed in the practical work of making drawings from the object it is really always done. The slates at right angles to each other are used simply to illustrate the principles governing the making and arrangement of the views, and in the work of the draughting office, as well as that of the elementary school-room, no further use is made of the planes, for shop and elementary drawings are not arranged, with reference to the edges of the planes, as are drawings in which projection methods are used

33. Teachers who prefer to consider the object as seen from one direction and turned to present the surfaces to be represented, must explain the arrangement of views desired, and also the fact that a point of the top view is in a vertical line from the same point in the front view, and that a point in a side view is in a horizontal line through the front view of the same point. Pupils may be told these facts and be taught to make their drawings in accordance, but the reasons for these facts are most easily understood when they are clearly illustrated by use of the slates

To present the principles as explained above will take little longer than to state only the facts regarding the desired positions of the views, when the principles are not explained, these positions are very likely to be forgotten

34. After the pupils understand the arrangement and nature of the views, mention of the slates, or planes of the drawing, is unnecessary, and the work is practically the drawing of different views of an object, which is so turned that these views may be seen and drawn in their proper relations to the front view.

If the subject is presented to pupils too young to understand what has been explained. the work must be simply dictation or copying. Much of the time now spent on working drawings in the lower grades of the public schools is wasted, for the pupils can do no more than copy ; instead of copying, they might spend their time profitably upon free-hand drawing, and, when old enough to reason, take up the study of working drawings, and thus not be obliged to copy.

35. **Making Working Drawings.** — When given any object for which drawings are to be made, the first question is, what position of the object shall be represented, and the next, what views will best show its construction and require the least time to make.

The front view of an object should represent its front surface, when it is in the position in which it is intended to be used. If the object has no portion which is the "front surface," or which is more important than the others, and has no special position, its position is immaterial.

Practical drawings give no more views than are needed to show all the construction, but while studying the subject, many views of simple objects may be made.

36. When possible, objects for study should be placed so that one principal surface will appear of its real shape, in either the front, the top, or the side view, and the pupils should begin with this view, or with the one concerning which they know the most. To work advantageously upon subjects at all complicated, it is necessary to begin a second view as soon as the first view represents all the parts which exhibit their real shapes in it. The parts which show their real shapes in the second view can be drawn after the second view has given all that the first view represents, and then the parts whose real shapes are seen in the second view may be projected to the first view. Circular parts should always be represented first in the view in which they appear as circles. By drawing vertical projecting lines from the front to the top or bottom views, or horizontals from the front to the side views, the different views will be made to agree with and to complete one another. In this way several different views can, and should be carried along together. In order to make them accurately and rapidly, the dimensions of circles, or of any parts which have the same size in two or more views, should be taken in the compasses or dividers and set off at once in all the views. This method is much more accurate and rapid than working by projecting lines, which often are not drawn quite parallel; this is the draughtsman's method, and though it may not be possible for young pupils to work in this way, it should be explained, and older pupils should use it.

37. Views of a Circular Plinth. — Fig. 8 is a perspective of a circular plinth of which working drawings are to be made.

FIG. 8.

If the plinth is a circular box it will be placed horizontally, if it is a clock it will be placed with its circles vertical. In general the object may occupy

any position, but it is better to have it placed in
the position in which it is intended to be used.
Suppose the circles to be vertical and the object
to be seen from the front, so that the front surface
appears of its real shape. The circle *F*, Fig. 9,
will then be the front view of the object. This
circle should be drawn before any of the lines of
the top view are drawn.

FIG. 9.

To see what the top view will represent, hold
the plinth so that its front face appears a circle, as in *F*, Fig. 9,
and then turn the top of the plinth toward the eye until the circles
appear as straight lines ; that is, as nearly straight as it is possible
for them to appear when held in the hand. In this position the
curved surface is seen, and the outline of the object is very nearly a
rectangle. The circles cannot appear as straight lines at the same
time, but the block can be held so that the circles, one at a time,
will appear straight; and, since the circles are perpendicular to the
top plane, this is the appearance which both must present in the top
view, *T*, Fig. 9. The length of the rectangle of the top view must be
the same as the diameter of the circle *F*, and its width must be the
same as the thickness of the plinth. The horizontal lines which
represent the two circles must then be as far apart as the plinth is
thick, and the short vertical lines must, if extended, be tangent to
the circle of the front view; they should be placed by drawing tangents (projecting lines) to the front view, or by setting off, with the
compasses, the required distance from either side of the centre line,
as explained in Art. 36.

38. Views of an Hexagonal Plinth. — Fig. 10 is a perspective
of an hexagonal plinth, or box, of which working drawings are to be
made from the object.

FIG. 10.

The plinth being horizontal when in use, the top
hexagon will appear its real shape in the top view.
This view should be drawn first, the plinth being
placed so that if any vertical faces are more important
than others they may be seen in the front and side
views of the object. This may cause the hexagon of the top view to
have a long diagonal horizontal or vertical.

Suppose the hexagon *T*, Fig. 11, to be the top view of the box,

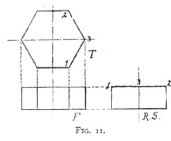

and the front and right side views to be required. To see what the front view will represent, the box must be looked at horizontally when placed as shown in the top view ; or, if we move the object to present the different views, the box should be held so that its hexagonal faces are vertical, each has two

Fig. 11.

edges horizontal, and so that only the hexagonal face, originally the top of the object, is visible. When in this position the lower horizontal edge of the visible hexagon represents the face of the plinth, originally its front vertical face. The object is now to be revolved so that this lower edge is brought upward toward the eye until the hexagons appear, as nearly as possible, straight lines. This position gives the appearance which the front view must represent; in this view three vertical faces are visible as surfaces, and the outline of the view is a rectangle.

Knowing the appearance which the front view must represent, it may be drawn in its proper position below the top view. The vertical lines of the front view are equal in length to the thickness of the plinth, and must be vertically under the corners of the hexagon which in the top view represent these lines. The horizontal lines which represent the hexagons are thus as long as a long diagonal of the hexagon. See *F*, Fig. 11.

To obtain the right side view, the object should be held to appear as shown in the front view, and then be revolved about a vertical axis, the right vertical edge coming toward the eye, until the object is seen at right angles to the direction in which the front view was visible. In this position the two visible vertical faces will appear of equal widths, the vertical surface originally at the front of the object will appear a vertical line at the left of the object, and the vertical surface originally at the back, a vertical line at the right of the object. The distance between these surfaces is the length of a short diagonal of the hexagon. These facts having been noted, the right side view may be drawn in its proper position at the right of and on the same level as the front view, its width being the distance 1–2 of the top view. See *R.S.*, Fig. 11.

39. Views of a Plinth and Disc. — Fig. 12 represents an equilateral triangular plinth support-ing, by means of two wires, a vertical circular disc, which appears a circle in the front view.

To represent this object only the front and side views are necessary; but the top view is added.

It is natural to draw first the triangular plinth, because it supports the disc. The disc appears a circle in the front view, and the plinth appears a triangle in the side view, so it is evident that the two views should be carried along together as ex-plained in Art. 36.

Fig. 12.

The side view *S*, Fig. 13, in which the plinth appears a triangle,

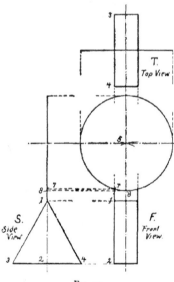

should first be drawn; but this view cannot be completed to rep-resent the circle until the front view of the circle is drawn; for the wires ½″ long, which support the circle, cause its lowest point to come below the ends of these wires an unknown distance.

Having completed the triangle of the side view, the front view of the plinth should be drawn. To discover its appearance, hold the plinth in the position represented by the side view, and then revolve the object on a vertical axis so that the right hand points of the plinth move toward the eye until the vertical triangles appear as nearly as possible vertical lines.

Fig. 13.

One sloping face is then visible; the horizontal base appears a hori-zontal line whose length is equal to the thickness of the plinth; and the triangles, one at a time, appear vertical lines, whose length is equal to the distance 1–2 of *S*. The front view of the plinth may now be drawn at the right of and on the same level with the side view. In this view the wires which support the disc are seen of their real lengths; they should be represented when the front view of the plinth is completed.

To obtain the circle which represents the disc, describe arcs, whose radius is that of the circle, from the upper ends of the wires. These arcs intersect at 5, the centre of the circle, from which the circle may be described. In the side view the circle appears a vertical line; this view may now be completed by limiting this line by a horizontal projecting line tangent to the circle at its highest point.

The top view of the plinth alone might have been given before this, but as there are no parts which must be represented first in the top view, the time of drawing it is not important. To see the appearance that the top view represents, hold the object as it appears in the front view, and then revolve the top toward the eye until the circular disc appears a horizontal line. In this position the length of the rectangle which represents the plinth will be equal to 3–4, or the length of the base of the object as seen in the side view.

The appearance having been discovered, the top view may be placed above and in line with the front view, by means of vertical projecting lines through the points of the front view. See *T*, Fig. 13.

40. Views of a Box and Pyramid. — Fig. 14 is a perspective of a tin box with its cover opened back and resting on the surface that supports the box. A square pyramid is centrally placed on the top of the box, the edges of its base being at 45° to those of the top of the box. The view which shows the front of the box, we will call the front view. This and the top and side views are required. The box is 6″ long, 4″ wide, and 3″ deep. The base of the pyramid is 3½″ square; its axis is 6″ long.

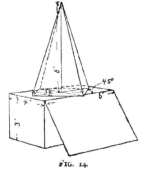

FIG. 14.

In the front view the box is represented by a horizontal rectangle 3″ high and 6″ long. This may be drawn first, and then the top and the end views of the box.

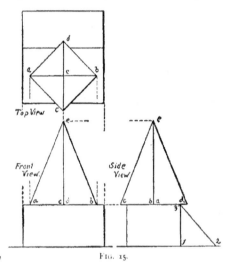

FIG. 15.

The top view of the box is a horizontal rectangle 4″ wide and 6″ long ; the side view is a horizontal rectangle 3″ high and 4″ long, and is on the same level as the front view, while the top view is directly over the front view. In the side view the cover is seen extending obliquely from the top of the box to the level of the bottom. To draw the side view of the cover, an arc of a circle whose radius is 4″ should be described from the edge, where the cover swings (point 3) to intersect the line of the table in 2. The cover must be represented by a line drawn from 2 to 3. The top view of the cover may now be drawn ; its width in this view is the distance 1–2 of the side view.

In the top view the square base of the pyramid is seen of its real shape ; this should be the first view of the pyramid drawn. To obtain the square, draw the diagonals of the top view of the box; these intersect at its centre. With this point as centre describe a circle 3½″ in diameter. Tangent to this circle and at 45° with the edges of the box, draw the sides of the square which represents the base of the pyramid. After this is drawn, the top view may be completed by drawing the diagonals of the square to represent the lateral edges of the pyramid.

In the front and side views the base of the pyramid coincides with the top of the box. In the front view the axis of the pyramid is a vertical line just under the centre of the square, — the point which represents the axis in the top view.

In the side view the axis is a vertical line midway between the front and back of the box, as is shown by the top view. In the front view the width of the base is given by projecting from *a* and *b* in the top view, and in the side view the width of the base must be equal to *c–d* of the top view. From *c* and *d* in the side view, and *a* and *b* in the front view, the contour lateral edges extend to the top of the axis. In both front and side views, the nearest lateral edge of the pyramid is a vertical line ; this line represents also the axis of the pyramid

41. It will be difficult to hold the models so that they have the proper relations and present to the eye the appearances of the different views of the complete group　The box and the pyramid may be held separately and turned to give the appearances of the different views of the single objects, as has been explained. It may happen that views of a group are desired when the relations are not so easy to see as in this group, in such cases it will be much better to arrange the group and to look at it in the directions of the different views required.

42. The making of simple working drawings in the manner explained, requires but slight knowledge of the principles of projection. The essential points are that the front and top views of the same point shall be in the same vertical line, and that the front and side views of any point shall be in one horizontal line. When the drawings are made from the object, they are simply representations of the appearances which it presents when seen from different directions, perspective effects of vanishing not being given. Pupils will readily learn to look at an object from different directions so as to realize the appearances which its different views will present; they will also learn to revolve the object to present these different appearances. Both methods should be explained, and pupils will then have little difficulty in using either

43. The simplest way to illustrate the manner in which an object should be turned to present the appearances which the different views are to represent, is to fold about the front, top, bottom, left, and right sides of a cube, squares of paper which form the development of these faces, and write front, top, bottom, left, and right sides respectively upon these different surfaces. If the cube is held so that the front view is visible, and the other marked sides occupy their proper positions, and if the squares upon them are then re-volved about the edges of the front face into the plane of this face, the manner in which the cube must be turned to present the different appearances will be understood at once, and also the relations of the different views to the front view. This experiment should be made by all the pupils

DIMENSIONING.

44. To dimension a drawing is to place upon it all measurements of the object represented, so that the workman may construct the object from the given measurements.

In order to be easily read the dimensions should be so placed as not to crowd or interfere with each other or with the lines of the drawing, and the figures must be neatly made, poor figures will spoil the appearance of the best drawing.

Dimensions should be put down in inches and fractions of inches up to two feet, thus $16\frac{3}{4}''$.

Distances greater than two feet should be given in feet and inches, thus $3' 7\frac{5}{4}''$.

Dimensions should be placed upon dimension lines which, by means of arrow-heads at each end, indicate the position of the dimension. These lines should not be continuous, a space being left to receive the figures, which should be symmetrically placed upon the line, thus. $\longleftarrow \quad 4\frac{1}{2}'' \quad \longrightarrow$
The line may be omitted when the space between the arrow-heads is short, and when there is not room for both arrow-heads and dimension, the arrow-heads may be turned in the direction of the measurement and placed outside the line, thus: $3\frac{3}{7}''$

When there is not room even for the dimension, arrow-heads may be used either outside or inside the dimension lines, and the dimension placed where there is room for it, thus. $2''$

Arrow-heads and figures should be drawn free-hand, when the drawing is inked they should be drawn with a common writing pen and with black ink, while the dimension lines should be red. The line separating the figures indicating the fraction must be parallel to the dimension line.

Vertical dimensions should be placed so as to read right-handed, thus :

The dimensions may be placed upon the drawing when there is room ; but when the space is small it is better to carry the dimension outside the drawing by means of dot and dash lines, thus :

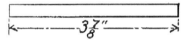

The space between the different views is often the best position for many of the dimensions.

When an object is divided into different parts and the lengths are given in detail, an over-all dimension should be given.

Dimensions should not be placed upon centre lines.

Distances between centres of all parts, such as rods, bolts, or any evenly spaced parts, should be given ; and when the parts are arranged in a circle, the diameter of the circle passing through their centres should be given.

The diameter of a circle and the radius of an arc should be given. The centre of an arc, when not otherwise shown, should be marked by a small circle placed about it, and used instead of an arrowhead. The dimension line should begin at this circle.

Dimensions should be clearly given in some one view, and not repeated in other views ; they should seldom be placed between a full and a dotted line, or between dotted lines when they can be placed in a view where the part is represented by full lines.

45. Simple objects, circular in section, are often shown by only one view. The fact that they are circular is indicated by *D.* or *Dia.*, understood to mean diameter, placed after the dimension.

46. When several pieces are alike, only one is drawn, and the number to be made is expressed by lettering.

47. When parts are to be fitted together, it is customary to write whether the fit is to be "tight" or "loose."

48. Surfaces which are to be finished by turning or planing are often indicated by placing upon a line drawn parallel to their outline, the letter *f*, which means finish.

49. Any special information which cannot be expressed by drawing is always expressed by lettering.

The name of the object represented, the scale of the drawing, its number and date of completion, should be placed upon the drawing.

Simple letters should be used for all this work ; they should be small and not more prominent than the drawing; pupils will find block letters made of a single line the best to use.

The draughtsman will find Soennecken's system of round writing neat and practical.[1]

The following are styles of lettering suitable for use :

ROUND WRITING.

[1] "This is not a system of lettering, but a scientifically evolved system of writing which has the effect of lettering without requiring the same amount of skill and consumption of time."

GEOMETRIC LETTERS.

A B C D E F G H I
J K L M N O P Q R
S T U V W X Y Z
a b c d e f g h i
j k l m n o p q r
s t u v w x y z
1 2 3 4 5 6 7 8 9 0

GOTHIC LETTERS.

A B C D E F G H I
J K L M N O P Q R
S T U V W X Y Z
a b c d e f g h i
j k l m n o p q r
s t u v w x y z
1 2 3 4 5 9 7 8 9 0

CHAPTER IV.

DEVELOPMENTS.

50. THE development of any object gives the real shapes of all the surfaces of the object. It is obtained by unrolling or unfolding the surface and placing it upon a single plane surface. Objects are developable or non-developable, according as their surfaces may or may not be laid out on a plane surface.

The cube, cone, and cylinder are types of developable objects; the sphere is a type of an undevelopable object.

51. The Cube. — The development of the cube, Fig. 16, may be

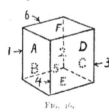

FIG. 16.

obtained as follows : Place any one of its faces, as *A*, upon paper, and with a sharp pencil trace its outline. Then tip the cube over, revolving it upon the edge *1* of the face *A*, until a second face, *B*, rests on the paper ; then trace face *B*. In the same way trace faces *C* and *D*, after revolving upon edges 2 and 3.

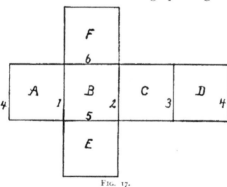

FIG. 17.

The development of these four faces is a rectangle whose width is equal to a side of the cube, and whose length is equal to four times the side of the cube. After the four faces *A*, *B*, *C*, and *D* have been placed upon the sheet of paper and drawn, revolve the cube so as to bring the faces *E* and *F* to the paper, and trace them ; these faces should be so placed that one edge of each coincides with an edge of the face *A*, *B*, *C*, or *D*.

The real shape of each face is thus placed upon paper in what is the most convenient way, supposing the cube is to be constructed of tin, paper, or any other thin material capable of being bent, for if the material is cut to the given outline and then bent at the lines *1, 2, 3, 4, 5*, and *6*, only the other edges of the object require fastening by glue or solder. See Fig. 17.

The development of an object is thus a pattern. All objects of sheet metal, from the simplest tin dish to the most complicated sheet-iron work in pipes, ventilators, and boats are obtained by means of such patterns.

52. When the form developed is to be constructed of paper, the outside edges of the development should be provided with projecting pieces called *laps*, by which the different parts may be held together. These laps are shown in some of the developments given in the plates of this book. Pupils making the drawings should construct the forms by developing, cutting, folding, and gluing the parts together, unless the objects are made in this way, many of the pupils will not understand the subject of developments.

The paper to be used for this work should be as stiff as a good quality of drawing paper. A medium weight manilla is satisfactory. All the edges to be folded may be cut partly through on one side by a knife, to cause them to fold sharply and neatly. The laps should be placed inside the object and fastened by glue or thick mucilage.

Manual and mental training of the greatest value are given by the making of objects which have been drawn and developed. This work has the advantage of being equally adapted for boys and girls, and of requiring no special or expensive materials or tools.

53. When these experiments have been made and understood, pupils will be able to apply the principles to the development of solids, represented by different views, without using the object; or, if the object is used to obtain the views, it will not be necessary to place it upon the paper and trace its different surfaces.

This method of tracing is, at best, an inaccurate one, its chief value is to illustrate principles so that pupils may understand that the development of any object gives the actual size and shape of every one of its plane surfaces, and the actual dimensions of all of its curved surfaces which are developable. After these points are understood, pupils will be able to make accurate developments by working from the different views of an object.

54. Prisms. The development of any prism will be obtained by placing its different faces upon the paper, and tracing their real forms upon it. The order in which these surfaces are drawn is of little consequence. The most natural way is to revolve the object in one direction, tracing the different faces, until all the lateral faces have been drawn. The bases may then be drawn so that a side of each coincides with an equal side of any one of the faces already developed.

55. The Cylinder. The development of the curved surface of

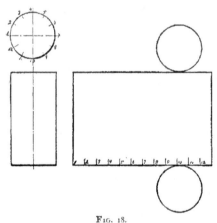

a cylinder will be obtained by rolling the cylinder along, and tracing as it moves, until its entire curved surface has passed over the paper.

A right cylinder will thus produce a rectangle whose width is equal to the length of the cylinder, and whose length is equal to the circumference of a base of the cylinder.

In exact work the circumference of the circular

Fɪɢ. 18.

base should be calculated, but in practice it may be obtained by dividing the circle into such a number of equal parts that the difference in the distances between two points in the circle measured in a straight line (the chord) and measured on the circle (the arc) is very slight.

56. By using the dividers and spacing the circle accurately, the study of working drawings may be carried on without the introduction of mathematics; if the points in the circle are not too far apart, this method is accurate enough for much practical work. The fact that this method is only approximate should be explained, and when the pupils are older and have had sufficient practice with instruments suitable for accurate work, they should calculate the circumference. This cannot be done before the pupils enter the high school, and not often then, for the instruments generally provided are such that perfect work cannot be obtained by any method.

57. In Fig. 18 the circle is divided into twelve parts, which for most work in the public schools, will be found better than a larger number, since the pupils do not work accurately enough to warrant smaller spaces on the circle.

If a good pair of spring dividers could be used and the entire circle accurately spaced, twenty-four parts would be preferable to twelve ; but the pupils will do all that can be expected, if they understand the principles and succeed in making models which are neat illustrations of the forms.

The bases of the cylinder are circles. In the development, they may be drawn tangent, at any points, to the sides of the rectangle which represent the development of the edges of the bases.

58. **The Cone.** The curved surface of the right cone, Fig. 19, will be developed by rolling the cone and tracing as it moves until all its convex surface has coincided with the plane of the paper.

If a straight line, $V-4$, drawn on the cone from the vertex to the base, is placed upon the paper, and the cone then rolled until this line returns to the paper, all the curved surface will have rolled over the paper. All the lines that may be drawn from the vertex to the base are the same length. As the cone rolls, it moves about its vertex V, which remains stationary. One after another the different lines, or elements, of its surface come

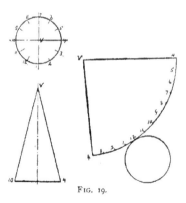

Fig. 19.

to the paper, and, as they are of equal length, the development of the circumference of the base is an arc of a circle whose centre is the vertex of the cone. The radius of this arc is the length of the straight line $V-4$, from the vertex to the base, and its length is equal to the circumference of the base of the cone.

To determine the length of the arc bounding the development of the lateral surface, divide the base of the cone into twelve equal parts, as in the case of the cylinder, and set off, upon the arc, one of these spaces twelve times.

The development of the base of the cone is a circle equal to the base. This may be drawn tangent to the arc at any desired point.

59. Pyramids. Fig. 20 represents a square pyramid, whose surface may be developed by placing any edge, as *1–V*, upon the paper

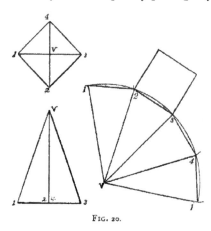

and revolving the object until its four triangular faces have been traced as they have coincided with the paper. The triangles are equal and isosceles; their bases form the sides of the base of the pyramid. The lateral edges of the pyramid are of equal length, and in the development, extend from the vertex *V* to equidistant points in an arc whose radius is the length of the lateral edges. The distance *1–2, 2–3, 3–4* and *4–1* in this

FIG. 20.

arc, is equal to the length of a side of the base of the pyramid.

The development of the square base of the pyramid may be placed so that one side of the square coincides with a base of any one of the triangles forming the lateral surface of the pyramid. In this way, any regular pyramid may be developed.

The distance from *V* to *1* or *3* will give the real length of the lateral edges only when the top view shows that these edges are parallel to the front plane.

60. The Sphere. The solids whose surfaces have been developed illustrate the way in which the development of any solid bounded by faces or by surfaces of one curvature may be obtained. The sphere is bounded by a surface which is curved in every direction; this surface cannot be laid out upon a plane, for the sphere touches a plane in one point only and as it rolls describes a line

upon the plane   The surface of the sphere is similar to those of
many other solids which are curved in more than one direction and
thus cannot be developed.

Though the surface of the sphere cannot be developed, a spherical or other surface of double curvature may be covered with paper,
metal, or other thin substance, by using many small pieces, each of
which stretches slightly in places and is compressed in other places,
so that practically they cover the surface.   The globe is thus
covered with narrow strips of paper which taper to a point in opposite directions.   Each of these strips is the development of the
curved surface of one of the equal surfaces of a solid bounded by
edges which are equal circles intersecting each other at two opposite
points, and by curved surfaces which are straight in one direction
and connect these edges.   These curved surfaces come within the
surface of the sphere, but if the surfaces are narrow they are within
that of the sphere only a very short distance.   In the case of a globe
covered in this way the surface of the sphere may be supposed to be
intersected at equal intervals by planes passing through a diameter.

The surface of the sphere may be intersected by parallel planes
perpendicular to a diameter of the sphere, and its surface covered by
the developments of the surfaces included between the cutting planes.
These planes intersect the surface in parallel circles, any two of
which may be supposed to be the bases of the frustum of a cone, if
the two circles are near each other the curved surface of the frustum
will come but slightly within that of the sphere and in practice
a spherical surface may be covered by assuming and developing the
surfaces of many frusta.

To obtain these developments the real dimensions of the assumed
surfaces must be determined by measuring the real length of each
line by which the real shape is found in the view in which it appears
its real length, or, if its real length cannot be seen in any view, by
finding it as explained elsewhere

Any surface which is curved in two or more directions may be
developed by assuming and developing cylindrical or conical surfaces which approximate the given surface.

It will often be necessary, instead of proceeding in any of the
ways explained, to divide a surface, by means of lines drawn upon it,

so that triangular surfaces, which together form a series of plane surfaces which approximate the given curved surface, may be supposed to pass through these lines.

In practical work, sheets of metal cut to form the development of the assumed plane surfaces will bend and stretch to form the curved surface. Such developments require advanced knowledge and are not explained in this book.

61. The simplest way to present the principles of this subject is to trace around the surface of an object which is placed upon paper as explained in this chapter.

When developments are made from drawings, it is not important how the surfaces are placed as long as, when folded together, they will form the object. The surfaces may be placed as they would be arranged by tracing around the object, or as they would be arranged by tracing the surfaces, one after the other, upon a transparent plane placed in front of the object. This method is in harmony with the arrangement of views adopted in this book, and therefore the developments in the plates are arranged in this way.

CHAPTER V.

SHADOW LINES

62. SHADOW lines are wider than the regular lines of the drawing, they have the same effect as cast shadows and relieve the projecting parts so as to produce an effect of perspective, even in projection, which is without perspective. They thus make drawings easier to read. If they cannot be applied so as to produce this result, they should not be drawn.

Some draughtsmen suppose the light to come from behind the left shoulder, in the direction of the diagonal of a cube which is so placed that the front view of the diagonal is a line at 45° extending downward to the right, and the top view of the diagonal is a line extending at 45° upward to the right. The shadow lines in this case will generally be the lower and right hand lines of the front view, and the upper and right hand lines of the top view.

Suppose a cube resting upon a corner, to be revolved about a vertical axis. As the cube revolves the shape of its cast shadow is continually changing, and, since this cast shadow is composed of the shadows of the different edges which separate the surfaces in light from those in shadow, it is evident that these lines are continually changing. This being the case, no rule or conventional method will enable one to apply shadow lines to any given position of the cube, so that they shall represent the edges which actually separate light from shadow. To find these edges we must obtain the cast shadow of the cube; for, except in very simple cases, it is impossible to say, without finding the cast shadows, which are the edges that separate the light from the shadow surfaces This is a very complicated problem, and hence shade-lining the edges that separate light from shadow is not suited to the requirements of practical work. Not only this, but since the edges separating light from shadow surfaces are continually changing as the object moves, the shadow lines when properly placed upon the drawing cannot convey, at a glance, information of much value.

63. Shadow lines will be of little value if they cannot be applied according to some system which does not require much time or knowledge to determine them, and which always represents the same facts of form in the same way.

Instead of determining the edges which really cast the shadows, some draughtsmen place shadow lines on the right hand and on the lower lines of the front view, and on the upper and right hand lines of the top view. Conventionally shaded in this way, the result is very different from that given by shading the actual shadow edges. As clearness is the only point sought, it is not wise to follow the assumed direction of the light even to this extent if a simpler and clearer method can be used. As already explained, the views may be supposed to be made upon one plane by turning the object as explained in Arts. 32 and 43. In this case the light will have one direction in all the views and the shadow lines will come in the same position in all views, and thus be much easier to determine than when they are the upper lines of the top view and the lower lines of the front view.

This being the case, it is the custom of most draughtsmen to treat all the views in the same way, as if the light came from behind and at the left, its direction being 45° downward to the right.

Assuming this direction in all the views, the lower and right hand outlines of all projecting parts will cast shadows, and should be shade-lined as illustrated in the following cases :

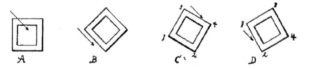

A square pipe in different positions is represented by *A*, *B*, *C*, and *D*. The direction of the light is represented by the arrow.

64. In *A*, the shade lines are the lower and right hand lines of the outside, and the opposite lines of the inside of the pipe. These lines at the inside are the upper and left hand lines of the drawing, but lower and right hand lines of the top and left sides of the pipe ; and every part of any object which is thus situated must be shade-lined as if it were a separate object.

CHAPTER VI.

INKING.

72. Pupils of the public schools, in any except advanced high school classes, should not attempt to finish drawings in ink, as they will obtain the best training and best results from the use of the pencil.

To ink accurately, it is imperative that an exact pencil drawing should first be made, its lines being fine and clear, and the centres of all arcs being carefully placed and marked by a small free-hand circle about them.

A special pen, called a *drawing pen*, and also special ink, are required to ink a drawing. The best pen for inking is one without a hinged joint, having its outer blade more curved than the inner one.

The illustration represents, full size, the lower portion of a pen suitable for students' use.

73. India Ink. The ink to be used may be liquid India ink which is sold in bottles, or ink prepared as it is needed, by grinding the solid stick India ink.

The liquid ink is prepared with chemicals which cause it to enter the paper so that its lines are erased with much more difficulty than those made from stick ink, which is ground as it is needed. Highly finished ink-drawings will be made more easily if the ink used is prepared by grinding, but the trouble of preparing it may render the use of the liquid ink advisable.

The stick ink may be prepared for use by grinding it in pure water, in an ink slab which may have either of the sections illustrated. The stick must be kept in motion all the time, slight pressure being applied, until the ink is thick enough to give, when dry, a perfectly black and solid line. After the ink is prepared, the stick should be carefully wiped to prevent its crumbling.

The slab must be kept covered to prevent evaporation and to keep the ink free from dust. If the ink hardens in the slab, it must be washed out. Fresh ink must be prepared every few days, as it spoils quickly.

74. Inking a Drawing. The pen may be filled with a quill toothpick. It is not necessary to move the blades to fill the pen; they should be set for the proper line and remain unchanged until all lines of that width are inked. When filled, the outside surfaces of the blades should be wiped with a cloth or chamois-skin to remove any ink upon them. If any remains upon that of the inside blade it will soon find its way to the straight edge, and then to the paper. The pen must be thoroughly cleaned after using, to prevent the blades from rusting

The pen should be held with the inside blade against the straight edge or French curve, and parallel to its edge. If it is turned so that the blades are at an angle, the ink will flow to the straight edge or curve and blot the paper. To keep the pen in the proper position, the forefinger may be placed upon the set screw, or the thumb may be placed on the edges at one side and the fingers on the edges at the opposite side of the blades.

The pen should be held with slight and even pressure against the straight edge or curve If the pressure varies, the blades will spring and the width of the line will change. Both blades of the pen should bear equally upon the paper A ragged line results when only one blade touches the paper. To correct this defect the pen must be inclined in the direction of the broken side of the line, until both blades bear equally upon the paper

The blades should be of such length that both will bear equally upon the paper when the pen is inclined slightly so as to bring the inner blade near the straight edge. If the pen does not make a good line when held as directed, it must be sharpened, but this cannot be done by young pupils.

The angle of the pen with the edge of the straight edge must not be changed while drawing any line, as this will vary the distance of the point from the straight edge and produce a crooked line

In going over a line the second time, the pen should be inclined and moved in the same direction as when the line was first drawn.

If a pen does not draw a smooth line without pressure, or if it cuts or scratches the paper, it should be sharpened. It will require sharpening often if it is used frequently, for the blades are quickly dulled by the paper

When inking circles or arcs, the pen point must be so inclined that both blades bear equally on the paper.

To ink very small circles a good bow pen is necessary, which, together with hair-spring dividers, will be required whenever really fine work is desired

All circles and arcs should be inked first, next the horizontal lines, beginning with those at the top of the sheet and working downward; then the vertical lines, beginning with those at the left of the sheet and working toward the right All parallel lines, when not horizontal, vertical, or at any of the angles given by the triangles, should be inked at one time, by means of triangles used as explained in Art. 6.

If the lines are not inked in order, from the top downward and from left to right, they will be blotted by placing the straight edge upon them before they are dry

To obtain perfectly tangential arcs, the tangent point must be found by drawing a pencil line connecting the centres from which

the arcs are drawn . to obtain perfectly tangential straight lines and arcs. the tangent points must be found by drawing a straight line from the centre of each arc, perpendicular to each straight line to which it is tangent. The draughtsman may work without these aids, but they are required by students.

Hatching lines should be a little finer than outlines, and should not be placed nearer together than is necessary to avoid the effect of a series of bars or wires, which will be given when they are too far apart. They should present the appearance of a tint . from ten to twenty lines per inch will give satisfactory results.

When many working lines radiate from a point, all should not be inked to the point, as this would form a blot ; they should stop at unequal distances from the point.

In inking a symmetrical curve, such as the ellipse, part of the curve each side of the axis should be struck with the compasses from a centre in the axis of the curve. This should be done, even if no more than $\frac{1}{8}$ or $\frac{1}{4}$ of an inch can be drawn in this way.

When the surface of the paper has been roughened by erasing, it may be made smooth by rubbing it with the clean, polished, rounded, ivory handle of a knife or other article.

75. Different materials are shown in inked drawings by using different colors for the line sectioning, or by tinting the sections with washes of different colors. Centre lines are generally inked full red; cast iron is sectioned black; wrought iron or steel, blue; brass, yellow; other materials are represented by other conventional methods.

76. Blue Prints. Shop drawings are generally blue prints. These are really photographs printed from a drawing made on tracing cloth. The lines are white on a blue ground.

When a drawing is finished wholly in black lines, or made for reproduction by the blue print process, different materials are shown by different hatchings. The materials commonly used are often represented as follows:

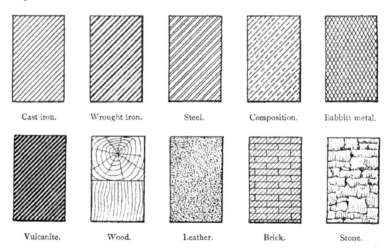

| Cast iron. | Wrought iron. | Steel. | Composition. | Babbitt metal. |
| Vulcanite. | Wood. | Leather. | Brick. | Stone. |

77. Erasing and Cleaning. The student should make erasures in an inked drawing wholly by the use of pencil or ink erasers, for when the knife is used for this purpose the surface of the paper is injured, so that the lines cannot be inked again neatly. The pencil eraser will, if used for several minutes, remove ink lines without injuring the surface of the best drawing papers. The ink eraser removes the lines more quickly, and generally gives satisfactory results.

When the inking is finished the whole drawing may be cleaned by rubbing it with bread, which is not greasy or so fresh as to stick to the paper. If the paper is much soiled it may be necessary to use an eraser. A soft pencil eraser should be used and great care taken that the ink lines are not lightened and broken by it

To avoid the necessity of using an eraser upon a finished drawing, instruments and paper must be kept free from dust and dirt. The triangles and T-square should be cleaned often, by rubbing them vigorously upon rough clean paper.

78. To Sharpen the Pen. — The blades of the pen should be curved at the points, and elliptical in shape. To sharpen the pen, screw the blades together and then move the pen back and forth upon a fine oil-stone, holding it in the position it should have when in use, but moving it so that the points are ground to the same length, and to an elliptical form. When this form has been secured, draw a folded piece of the finest emery paper two or three times between the blades, which are pressed together by the screw. This will remove any roughness from the inner surfaces of the blades; these surfaces should not be ground upon the oil-stone

When the blades are ground to the proper shape, they must be placed flat upon the stone and ground as thin as possible without giving them a cutting edge. To do this, the pen should be moved back and forth and slightly revolved at the same time. Both blades must be made of equal thickness. If either blade is ground too thin, it will cut the paper as would a knife, and the process must be repeated from the beginning. In order to see the condition of the blades, they should be slightly separated while being brought to the proper thickness.

79. Stretching Paper. — When paper is to be stretched it should be dampened and then immediately secured to the drawing board by mucilage It is not necessary to strain the paper so that it is flat while it is wet, it is better not to do this, for, if this is done, the tension created in drying may cause the paper to tear if the board is dropped, and when the paper is cut from the board, it may shrink enough to change the dimensions of the drawing a sixteenth or, if the drawing is large, even an eighth of an inch.

80. The following are the steps involved in stretching paper

1. Turn over and press down for about $\frac{1}{2}''$ the edges of the paper, all the way around the sheet

2. With a clean sponge and water, dampen all the upper surface of the paper, including that of the folded edges   Allow the sheet to stand, and moisten again if necessary, until the paper has become dampened and swollen throughout. Time may be saved by moistening both sides of the paper. When this is done the side of the folded edge which is to receive the mucilage should be dried with a blotter or cloth.   The paper should not be rubbed with the sponge, as this will roughen and destroy the surface.

3. Apply to the folded edge a thick mucilage made by dissolving cheap gum arabic in cold water.

4. Turn the edges over and press them upon the board, beginning in the centre of each side and working toward the corners

5. Press the edges upon the board and rub their surfaces until they hold firmly.

6. See that the mucilage holds firmly before leaving the paper to dry, as the strain will pull the paper from the board if the mucilage does not set while the paper is damp

7. The board should be left in a horizontal position with no water upon the surface of the paper, it should never be placed near a radiator or in the sun.

8. The edges may be quickly set by rubbing them with a piece of polished hot metal.

   Mucilage is better than glue for stretching paper, as it does not set too quickly, and the edges of the paper may be easily removed from the board when the sheet is cut off To remove these edges, cover them with water, after a few minutes the paper will absorb the water and they can readily be removed

CHAPTER VII.

MACHINE SKETCHING AND DRAWING.

81. PROJECTION forms the basis of practical working drawings, in order to make technical drawings one who understands projection has simply to become familiar with the different conventional representations and methods which cause practical drawings to differ from complete projections of the objects

The draughtsman finds it necessary to make drawings of machines already made, as well as of those which he designs. To do this he makes free-hand sketches of the machine, measures all its parts, and places the dimensions upon the sketches, never making more views than are necessary to show the construction of the object represented. He then returns to the office and makes, from his sketches, finished drawings to scale

This book is intended to explain projection principles and also the making and arrangement of practical working drawings, and illustrates drawings of both kinds The objects used by pupils for study must be simple and easy to obtain In order that pupils may understand how to represent more difficult subjects, it is necessary to give more views than would be required to make the objects. This should be remembered when the plates are studied.

Projecting, or guiding, lines as they are sometimes called, are not given in practical drawings, and are given in this book in the drawings of machine details only when they are necessary to show how projection principles are applied to special points.

In order to make complete working drawings of the drawings previously made and explained, it will only be necessary to place dimensions upon them and to indicate the material to be used and the amount of finish to be given the different parts.

In the plates dimensions or dimension lines have been placed upon a sufficient number of drawings to illustrate the way in which dimensions should be placed.

82. Sections are of great value to the machine draughtsman, as they enable him to dispense with many complicated views They are usually taken horizontally or vertically, but may be taken in any direction , they are often taken at right angles to parts whose real shapes are not shown by views of the outside of the object. The part of the object behind the cutting plane is generally shown in addition to the surface cut by the plane , but the surface cut by the plane is often all that is necessary to give the desired information, and is all that is given in many drawings representing sections. The section of the arm on line *AB* in Fig. 117 illustrates this point. Such a section may be placed as illustrated It is often found about *AB* and on the part sectioned ; in this case the part is often represented as broken, thus leaving a space for the section.

83. When an object is too long to be shown of its entire length, and is of one shape throughout or for any great distance, the central part may be considered removed, each end being represented as broken, and the entire length shown by the arrow-heads and figures.

The positions at which sections are taken are usually indicated by dot-and-dash lines

84. The surface of the material cut by the cutting plane is cross-hatched with lines which are generally drawn at $45°$, and in the same direction on the same piece, wherever it may extend or however much it may be cut up or intersected by other pieces The distance between the lines must be the same throughout all the surface of any one piece.

When different pieces are cut by the plane, they should be hatched in opposite directions. especially if there is no space between them. When three or more pieces are cut and come together, they can be distinguished by a difference in the spaces between the hatching lines.

85. When objects such as bolts, rods, or other solid parts lie in the plane in which a section is taken, they should be shown in elevation : for time is required to hatch the parts, and nothing is gained by doing this, since they are solid. Fig. 125 illustrates this point.

86. It is not necessary that the section be taken on one straight line, it may be taken so as to produce the clearest drawing. Thus in Fig. 119 a vertical section through the upper part of the caster would cut the frame so as to show part in elevation and part in section. It is much clearer to suppose the cutting plane to pass obliquely through the centre of the frame, and to suppose the wheel to be cut by a vertical plane, than to adhere to the facts of projection given by the use of any one plane.

87. If a wheel having arms is situated with one or two arms in the plane of the section, neither arm should be sectioned, for the sectional view would then not differ from that of a solid wheel. Thus in Fig. 116 a horizontal section of the wheel would cut two arms, but should be represented the same as the vertical section

88. The aim of the draughtsman is to convey a clear idea of the object to be constructed, and, as certain parts and details which are common to various classes of work are generally perfectly known and always recognized from even one view, he often shows a part (for instance a set screw) in position in one view, and omits it in other views where simply its position is shown. Instead of perfect projections he is content to make conventional drawings , he is generally considered free to use his judgment, both as to the views which shall be made and as to the essentials of these different views , he never makes more views of an object than are needed to show its construction

89. It is necessary to make drawings which show all parts of an object in their proper positions. These views are called *assembly drawings,* upon them are placed important over-all dimensions, distances between centres, etc. For the workmen, detail drawings are prepared which give as many views as are necessary to show fully all the different parts which make up the complete machine.

The assembly drawing shows these details in the positions they occupy when in use, and many of them, owing to the positions they occupy, may not be clearly shown   The detail drawings arrange these parts, without regard to the position which any part occupies in the machine, so that their forms are clearest shown by the fewest views.

Other points are explained on the pages opposite the plates, Chap. XII.

CHAPTER VIII.

ORTHOGRAPHIC PROJECTION.

90. ORTHOGRAPHIC PROJECTION is the art of representing an object by means of projections or views made upon different planes, at right angles to each other, by the use of projecting lines perpendicular to the different planes and passing through all the points of the object

The work of the draughting office often requires no thought of the planes of projection, hence some teachers claim that reference to these planes is unnecessary, they assert that pupils should not be asked to consider problems of projection which are difficult for most of them to understand

It is easier to remember rules than to understand and apply principles. Working drawings may be made by rule with far less effort on the part of the teacher and pupil than when they are made as the result of knowledge of projection principles ; but this does not prove the principles useless or unnecessary

To draw from the object is very simple, and is all that can be expected from grammar school pupils; sometimes little more can be done in high, evening drawing, and in elementary technical schools; but in the high and elementary technical schools, there may often be found students able to reason and to understand the principles of projection well enough to draw the views of a simple object from a written description of the object and its position   Such students should have the benefit of the training given by a course in projection.   Many pupils unable to understand projection sufficiently to draw even simple objects from written statements, will nevertheless be interested and benefited by a talk upon its principles, and teachers should give instruction in the principles of projection whenever pupils are able to profit by such instruction.

Any draughtsman ought to be able to draw without the object before him all the time, or without being given in whole, or in part, one or two views of the object.   Yet the instruction given by many teachers, while it enables pupils to draw from the object, to complete unfinished views, and to add other views when some are given, does not give the capacity for representing objects conceived in the mind in definite positions

91. A common method of instruction is for teachers to place upon the blackboard two views of an object and ask the pupils to draw the third, or to give the axis or a line or two of drawings which are to be completed by the pupils   This method causes the students to remember the problem, or a similar one, and the steps involved in its working   There are few who with this assistance are unable to make the required drawings

If the representation of constructed objects was the only work of the draughtsman these methods would always be as satisfactory as they are in the simple work of object sketching and draughting, whether from the object constructed or imagined, but the draughtsman often has problems which require objects to be placed in certain definite relations to each other   In drawing, he must refer to something, and whatever this is called, it is the plane of the drawing or a parallel plane.   If he is unable to draw a line which makes definite angles with the ground and with the vertical plane of the drawing, he cannot represent an object as simple as a cube or cylinder, which is to have some definite position

Whatever may be done with the most elementary work, the advanced student should be able to draw simple geometric solids from a description of the solid and its position.   To do this the planes must be thought of ; and teachers who are proclaiming against the use of projection methods will do great harm if they confine students old enough to understand and profit by these methods, to work from the object, and they must be thus confined when projection methods are not used.

Models and objects are necessary in the first study of working drawings or of projection, but they should be dispensed with as soon as possible.

Pupils should learn to see the object mentally when its dimensions and relations to the planes are given, and to see also the projections of the object upon these planes.   They will then be able to draw without reference to models. Only in this way is it possible to understand the subject of projection; and it should be realized that those who are protesting against such study cannot produce a substitute method which will accomplish the same results.

92. When pupils are taught to make working drawings solely by observation of the object, it is impossible for them to work in any other way, they cannot draw even two views of the simplest form, such as a cone, whose axis makes an angle of 30° with the ground and 45° with the front vertical plane.

In giving an examination in the subject it is impossible, unless reference is made to the planes of projection, to call for two views of any line or object inclined to both planes, without giving lines or views which tell the pupils what they are to do, so that the work becomes memory instead of reasoning

The objection to reference to the planes of projection is absurd when coming from those teachers who give to their classes difficult problems of intersections and oblique projections, which cannot be solved except by the use of cutting and other auxiliary planes, whose relations are much more difficult to understand than those of the planes of projection. Though not named, the planes of projection must still be practically used, hence, they should be explained and the study carried on by projection methods when pupils attempt these problems.

93. Teachers of drawing in any school above the grammar, ought to understand projection and descriptive geometry, even if these subjects are not explained to their classes, they will then be able to answer questions relating to the representation of form, abstract or concrete, even if they have not studied the various methods in which the principles are applied to the practical work of the mechanic

It is true that knowledge of descriptive geometry will not enable one to represent details of construction in the conventional ways peculiar to the architect or the machine draughtsman, but this knowledge is the foundation for all drawings, and conventionalities are quickly understood and applied by one who is able to represent correctly any form in any desired position.

94. The principles of projection which have been explained in

connection with the study of working drawings cover the simple
work of the draughting office, and are all that the teacher in the
grammar grades requires , but teachers of drawing, advanced students in the high, and students in technical schools, should carry the
study farther. For their use the following drawings and explanations are given, and will be all that are required for knowledge of
the simple principles of orthographic projection. A thorough understanding of the subject requires a course in descriptive geometry.

95. This book is intended for teachers of public and elementary
schools, in which the subject of working drawings is more important
than projection   The best arrangement of views for working drawings is explained in Chapter III.   This arrangement is different
from that given by the planes generally used in the study of projection.

The drawings of the plates of this book are arranged as working
drawings should be arranged, and, in order that there may be no
confusion, the following notes on projection make use of planes of
projection which produce the arrangement of views chosen for use
in the study of working drawings

Some of the objects represented on the plates, with their relations to assumed planes of projection, are described at the end
of this chapter. These planes are not represented in the drawings,
and those who study the following notes may test their knowledge of
the subject by using the statements as test questions, and making
the projections required to represent the objects described

96. Projection Principles. — The front, top, side, back, and
bottom planes of projection are represented in Fig. 21.

The drawings or tracings made upon these planes are the different *projections*, or *views*, of the object. The perpendiculars to these
planes, by means of which the views are obtained, are the *projecting
lines*. The projections will be called views throughout this chapter.

In all the drawings a point will be designated by a letter or figure ; its projections by the same letter or figure with that letter as
exponent which represents the plane upon which the view is found.
Thus T means the top, F the front, R the right side, L the left side,
B the back, and G the bottom view   To avoid confusion the exponents are often placed upon only part of the letters; but in any view

where an exponent is found upon any letter, the same exponent should be understood when reading the other letters

97. Axes of Projection. Fig. 22 represents the top, bottom, and side planes when they have been revolved to coincide with the plane of the front vertical plane.

The lines in which the planes of projection intersect are called *axes of projection*. The different axes are named in Figs. 21 and 22.

When the planes are in the positions shown in Fig. 21, and the surface of any one plane is seen, the other planes are seen edgewise and are represented by the axes of projection. Thus, when the front plane is seen, the front horizontal axis at the top represents the top horizontal plane, the front horizontal axis at the bottom represents the bottom plane, and the right and left side planes are represented by the right and left vertical axes.

In the first study of projection the front vertical plane, the top horizontal plane, and the two side vertical planes are the only planes of projection which are generally required. Their axes of projection will be designated by the following abbreviations $F. H. Axis$, $R. H. Axis$, $L. H. Axis$, $R. V. Axis$ and $L. V. Axis$. The lower horizontal plane not being used, the terms are understood to refer to the upper axes.

98. Views of a Point. Fig. 21 represents a point, A, situated between the planes of projection. A projecting line from A to the front plane gives A^F, which is the front view of point A, a projecting line from A to the left side plane gives A^L, which is the left side view of A, a projecting line from A to the top plane gives A^T, the top view of A, projecting lines to the right side plane, to the back plane, and to the lower plane will give the views of A upon these planes. These views may be designated by A^R, A^B, and A^G respectively.

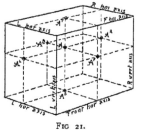

FIG 21.

99. The projecting lines from A to the vertical planes are horizontal and in a horizontal plane passing through A. This plane intersects the different vertical planes in four horizontal lines, which are at equal distances from the top plane, and form a rectangle.

When the planes of projection are revolved into the plane of the
front plane, these horizontal lines form one straight line which
passes through A^F and contains the different vertical views of A.

The projecting lines from A to the front, back, top, and bottom
planes are in a vertical plane which is perpendicular to the front
plane and parallel to the side planes. This vertical plane intersects
the front, back, top, and bottom planes in a rectangle whose sides
develop into a vertical line when the planes are revolved into the
plane of the front plane. Thus the views A^T, A^B, and A^G are in a
vertical line passing through A^F.

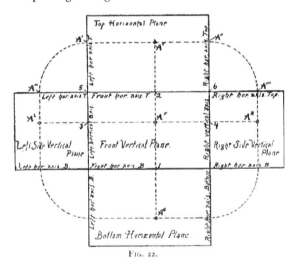

FIG. 22.

100. Position of a Point (Figs. 21 and 22). Point A is a certain distance from each of the planes of projection. In the front
view A^F–1 is the distance of A above the bottom plane; A^F–2 is the
distance of A below the top horizontal plane; A^F–3 and A^F–4 are the
distances of A from the side vertical planes. In the top view the relations of A to the side vertical planes are seen, also the distance
A^T–2, of A behind the front vertical plane. Projecting lines from
A^T to the right and left horizontal axes give the top projections A',
A'' of A upon the side vertical planes. When the left and right
horizontal axes are revolved about points 5 and 6 respectively, points

A''' and A'''' are obtained The side views of A are in verticals from these points and in a horizontal line through A^F. The distance A^L–$_3$ and A^R–$_4$ is thus the same as A^T–$_2$, and shows the distance of A from the front vertical plane. In the same way the view of A upon the bottom plane agrees with the other views.

101. Having the front and top views of any object or point given, the side views can always be obtained by projecting from the top view to the left and right horizontal axes, revolving these axes until they coincide with the front horizontal axis, then drawing verticals from the points in the horizontal axes and intersecting them by horizontals from the points of the front view. Having the front and side views given, the reverse of the above process will give the top or bottom view. When the planes are seen from above, and the points of the top view have been projected to the side planes (that is to the right and left horizontal axes) these projected points describe, when the planes (axes) revolve, arcs of circles whose centres are in the ends of the axes ($_5$ or $_6$, Fig. 22.) In the same way, the points of either side view, when they have been projected to the horizontal axis, describe, when the axis revolves, arcs of circles whose centre is an end of the axis.

102. In order to understand the subject, students must accustom themselves to looking at the different views separately When looking at A^F only the front view should be seen. The lower edge of the front plane should represent to their minds the bottom plane, which seen edgewise appears a line, the upper edge of the front plane should represent the top plane, of which the edge only is seen, and the vertical edges of the front plane should represent the vertical side planes, which seen edgewise, appear vertical lines.

In the same way, when the top plane or a side plane is seen, the edges of these planes should represent the planes at right angles to them. It is possible to memorize rules and methods and in this way to make drawings which are correct, but this does not give the power to do original work, or the best work, or to attain real understanding of the subject The lines must represent planes and solids and space, and only by means of this mental construction of the actual conditions, can drawings mean more than the lines of a complicated plane geometrical construction

103. Views of a Straight Line. Fig. 23 represents the planes of projection and a triangular prism formed by bisecting a cube as explained in Art. 27. The edges of this object represent the lines to be studied.

The views are obtained by means of projecting lines, which are perpendicular to the planes. It follows that a line which is perpen-

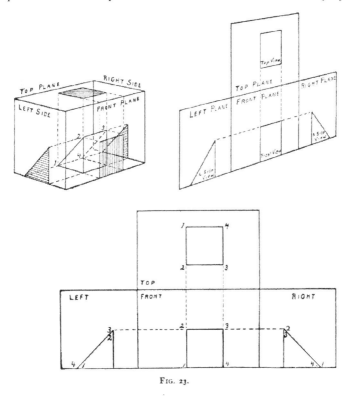

FIG. 23.

dicular to any plane will be projected upon that plane as a point, for the projecting lines from its two ends coincide with each other and with the line. Thus points *2* and *3* are the upper ends of vertical edges which are perpendicular to the top plane and are represented in the top view by points *2* and *3;* points *1* and *4* are the back ends of edges which are perpendicular to the front plane and are projected upon it as points *1* and *4*. Edges *2–3* and *1–4* are perpendicular to the side planes and are represented in the side views by points *2, 3* and *1, 4*.

The projecting lines to any one plane are parallel to each other; the distance between them must be measured perpendicularly, that is by a line which is parallel to the plane to which the projecting lines extend. It follows that the view of a line on a plane to which it is parallel must be parallel and equal to the line. Edges *2–3* and *1–4* are parallel to both top and front planes and their views upon these planes are thus parallel and equal to each other and to the lines.

Edge *1–2* is oblique to the top and front planes, therefore the projecting lines from its ends to these planes make its views shorter than its actual length. The edge *1–2* is parallel to the side planes, so its views upon these planes give its real length.

A line connecting the corners *1* and *3* is a diagonal of the sloping face of the prism; it is oblique to all the planes and therefore in all the views the distance *1–3* is less than the actual distance between the points 1 and 3.

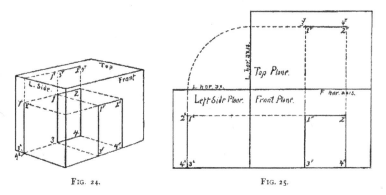

Fig. 24.   Fig. 25.

104. Views of a Plane Surface. Figs. 24 and 25 represent three planes of projection, and a rectangular card, 1, 2, 3, 4, parallel to the front vertical plane and perpendicular to the top and side planes. Those who understand the previous figures will understand these drawings at a glance. The front view gives the real dimensions of the card and the positions of its edges with reference to the top and side planes ; the top view gives the width of the card, its relation to the front and side planes, and its distances from these planes : the side view gives the length of the card, its relation to the front and top planes, and its distances from these planes.

105. Views of a Solid. — Figs. 26 and 27 represent a rectangular pyramid whose axis is vertical and at given distances from the planes of projection. Two edges of the base of the pyramid are parallel and two are perpendicular to the front plane.

Fig. 26 is a perspective view which represents the pyramid and the planes of projection, with the views of the pyramid upon them.

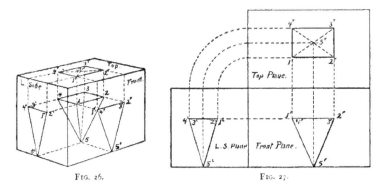

FIG. 26.   FIG. 27.

Fig. 27 represents the planes and their respective views when the top and side planes have been revolved into the plane of the front plane.

The base of the pyramid is at right angles to the front plane, upon which it is projected as a horizontal line I^F–2^F ; it is parallel to the top plane, upon which its real shape is given.

The axis, being perpendicular to the top plane, is projected upon it as a point 5^T, at the centre of the base, and to this point the top views of the lateral edges extend and form the diagonals of the figure. The right and left triangular faces of the pyramid are perpendicular to the front plane, and are projected upon it as lines 2^F–5^F and I^F–5^F. The two wide triangular faces of the pyramid are oblique to the front plane, and the two narrow triangular faces are oblique to the side plane ; thus the real shapes of these faces are not given in either view.

106. Simple geometric forms are used in the first study of projection. Generally the views of these forms upon the front, top, and one side plane are all that are required to describe the object. It is not customary to limit these planes of projection except by the three lines, or axes, in which they intersect. These axes are the front horizontal axis, a left or right horizontal axis, and a vertical axis at the left or right, and are represented in the illustration.

The *F. hor. axis* is the only one required to locate objects with reference to the front and top planes. Figs. 28, 29, and 30 show how the views of objects may be obtained, using only this axis.

107. Views of a Rectangular Card. — Fig. 28 gives the front and top views of a rectangular card, $2'' \times 4''$, which is parallel to, and

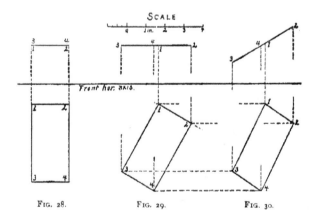

Fig. 28.   Fig. 29.   Fig. 30.

$2''$ behind the front plane ; its short edges are parallel to the top plane, the upper one being $1''$ below the plane.

The front view is a rectangle $2'' \times 4''$, its long sides vertical, and its upper short side $1''$ below the axis of projection. The top view is a horizontal line $2''$ above the axis of projection.

108. Fig. 29 represents the same card as Fig. 28, when it is parallel to the front plane, and the same distance behind it as in Fig. 28, but with its long edges at 60° to the top plane. The front view gives the real shape of the card, and the real angles which its edges make with the top plane. The top view is a horizontal line.

109. Fig. 30 represents the same card when perpendicular to the top plane, and the same distance from it as in Fig. 29, its edges are at 60° and 30° to the top plane as in Fig 29, but its surface is at 30° to the front plane, instead of parallel to the plane as in Fig 29.

When the card is in the position shown by Fig. 29, its top view is a horizontal line whose length is obtained by projecting from *2* and *3* of the front view. If the card is revolved about a vertical axis passing through point *1*, the angles of its edges and its surface with the top plane not changing, the top view will show the angle of the card with the front plane, but its length will not change, therefore, for all positions of the card during a complete revolution, the only change in the top view is in the angle which it makes with the front horizontal axis.

Suppose the card to revolve about point *1* as described, points *2, 3,* and *4* describe horizontal circles whose centres are in a vertical line passing through *1*. These circles appear circles in the top view, and as horizontal lines in the front view. It will be seen that points *2, 3,* and *4* of the front view, which represents the card when its surface is at an angle to the front plane, must be in horizontal lines drawn through the corresponding points of Fig. 29 Hence, to obtain the front view of the card when at any given angle, it is necessary simply to place the top view of Fig. 29 at the required angle and draw vertical projecting lines from all its points, and intersect them by horizontals from the corners of the front view of Fig. 29

The views of Fig. 30 cannot be obtained without making use of those of Fig. 29, or of substitute drawings The views of a solid oblique to one or both the principal planes of projection may be determined in the same way as those of the card

110. Views of a Pyramid. — Fig 31 represents a pyramid

whose base is a rectangle $2'' \times 3''$; its axis is $5''$ long, is vertical, and $2''$ behind the front plane; the vertex of the pyramid is $1''$ from the top plane, and the long edges of the base are parallel to the front plane.

III. Fig. 32 represents the same pyramid after it has been revolved to the right through an angle of $45°$, about the right edge of the base; all the edges of the object are situated with reference to the front plane as they are in Fig. 31.

Suppose the object to be moved to the right of its position in Fig. 31, its distance from the front plane not changing, and then

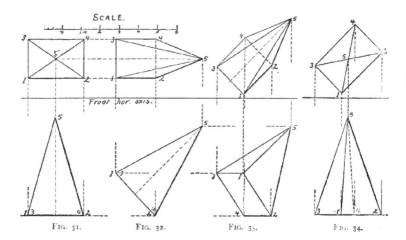

FIG. 31.   FIG. 32.   FIG. 33.   FIG. 34.

revolved to the right about the edge $2-4$. As the object revolves, all points in it move in arcs of circles whose centres are in $2-4$ and which are parallel to the front plane and appear horizontal lines in the top view. The points in the top view of Fig. 32 must then be in horizontal lines drawn through the corresponding points of the top view of Fig. 31.

The object may revolve through a complete circle, and cause no change in the size or shape of the front view; thus the front view of Fig. 32 will be the front view of Fig. 31, with the axis at an angle of $45°$ instead of vertical. The top view will, however, present an infinite number of different appearances, as the object revolves.

To obtain the views of the object when its axis is parallel to the front plane as in Fig. 31, and at any angle to the top plane, it is simply necessary to place the front view of Fig 31 with its axis at the required angle to draw vertical projecting lines from all its corners and intersect these verticals by horizontals from the corresponding points of the top view of Fig. 31. The front view in Fig. 32 might be drawn without reference to Fig. 31, but it will often be necessary to draw the views of an object situated as in Fig. 31, in order that the shape of the front view for Fig 32 may be known

112. Fig 33 represents the object when its axis is at an angle with both planes

Suppose the pyramid placed as in Fig. 32, with its axis at 45° to the top plane, to be moved horizontally to the right and then revolved about a vertical axis passing through its vertex, through a complete circle, the angle of the axis with the top plane not changing, and the distances of all points of the object from the top plane remaining as in Fig. 32 The top view of the pyramid in any and all of its positions must be the same in form and size as the top view in Fig. 32 Hence, to obtain the projections of the object when its axis is at 45° to the top plane, as in Fig. 32, and the vertical plane containing the axis is at any given angle, say 45°, to the front plane, we have simply to place the top view of Fig. 32 so that the axis of the pyramid makes an angle of 45° to the front axis of projection, and draw verticals from the points of this view, and intersect them by horizontals from the corresponding points of the front view of Fig. 32. Thus the front view of any point, as *r*, will be in a vertical from *r* in the top view of Fig. 33, and in a horizontal from *r* of the front view of Fig. 32.

In this way, by considering one point at a time, the views of the most complicated objects can be obtained.

Similarly, the front view of Fig. 33 may be revolved through any desired angle and a top view to correspond be found by projecting from the top view of Fig. 33, this process of revolving first one view and then the other may be repeated as many times as is desired

Generally no more than the revolution of one top view and one front view will be necessary to give views of any object at angles to both planes of projection

Fig. 34 shows the top view of Fig. 31 revolved through an angle of 45°. The front view in Fig. 34 may be revolved as desired and thus any appearance of the object be obtained

113. All the vertical and horizontal projecting lines may be drawn before any of the points of the required view are determined, but it is much simpler for elementary pupils to draw the projecting lines for the different points, one at a time, and mark the points of the view as they are determined.

The points of solids studied should be numbered, and the numbers carefully placed in all the views so that they indicate corresponding points. When this has been done, to obtain the points for any required view, as the top view of Fig. 32, it is simply necessary to draw a vertical from any point, as 5, of the front view and intersect it by a horizontal from 5 of the top view of Fig 31.

114. True Length and Position of a Straight Line. The true length of a line is given by any view only when the line is parallel to the plane upon which the view is made ; when thus situated the view gives not only the true length of the line, but its true positions with reference to the planes at right angles to the one upon which the view is made. Frequently a line is oblique to all the planes, and its real length is not shown by any view, hence it is important that a simple method of determining the real length of à line be found.

This problem is solved in the figures of this book as follows

Drawing A represents a line $a\,b$ which is situated so that its front view makes an angle of 30° with the top plane (F. H. axis), and its top view makes an angle of 45° with the front plane (F H. axis) To find the real length of $a\,b$ and the angle it makes with the top plane, the line must be revolved until it is parallel to the front plane, when it will be represented in the top view by $A^F\,b'$ As $a\,b$ revolves, b moves in an arc which, in the top view, is represented by the arc $b^T\,b'$, and in the front view by the horizontal line $b^F\,b''$. Line $a^F\,b''$ is the real length of $a\,b$, and gives the real angle of $a\,b$ with the top plane.

If the angle of $a\,b$ with the front plane is desired, $a\,b$ must be revolved until it is parallel to the top plane, as illustrated by drawing B. As $a\,b$ revolves, b moves in an arc which is represented in the front view by the arc $b^F\,b'$; and in the top view by the horizontal

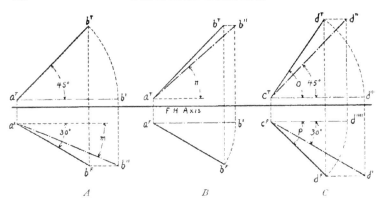

A   *B*   *C*

line $b^T b''$. Line $a^T b''$ is the real length of $a b$, and gives its real angle with the front plane.

In drawing A, the real length of a b *is equal to the hypotenuse of a right-angled triangle, of which the base is equal to the length of the top view of* a b, *and the altitude is equal to the difference in the distances of points* a *and* b *from the top plane.*

In drawing B, the real length of a b *is equal to the hypotenuse of a right-angled triangle whose base is equal to the length of the front view of* a b, *and whose altitude is equal to the difference in the distances of points* a *and* b *from the front plane.*

The real lengths of a number of lines of different lengths will be obtained most quickly by constructing, at any convenient place outside the drawing, right-angled triangles according to the above paragraphs; these triangles may have a common right angle.

115. When, instead of both views being given and the true length and angles being required, the angle of one view, and the real length of the line, and its real angle with the other plane are given, the views of the line will be found as follows:

Suppose that $a b$ is the length $a^F b''$ (drawing A) and makes the angle m with the top plane, while its top view makes the angle $45°$ with the front plane. Draw first $a^F b''$ of the proper length, and at the angle m with the top plane; then project and obtain $a^T b'$, which is the length of the top view. From a^T draw the top view at the given angle, $45°$, and make its length equal to $a^T b'$ by drawing the arc $b' b^T$. Project from b^T to the level of b'' and obtain b^F, join b^F and a^F and the front view is given.

The views will be obtained in the same way if ab is the length $a^T b''$ (drawing B), and at the angle n with the front plane, and having its front view at the angle 30° with the top plane.

116. Drawing C illustrates the way in which the views of a line are found, when, instead of the angles of the views, the real angles with both planes and the real length of the line are given

Line cd is the distance $c^F d'$ in length, it makes an angle of 30° with the top plane, and one of 45° with the front plane. To obtain the views, whose angles are not known, draw first $c^F d'$ equal in length to cd and at the angle 30° with the top plane, then draw $c^T d''$ equal in length to cd, and at the angle 45° with the front plane. From $c^F d'$ project to obtain $c^T d'''$, which is the real length of the top view of cd when it is at the angle 30° with the top plane. From $c^T d''$ project to obtain $c^T d''''$, which is the length of the front view of cd when it is at the angle 45° with the front plane. Point d must be as far from the top plane as is d', and as far from the front plane as is d'', hence, d^F must be in an arc whose radius is $c^F d''''$, and in a horizontal line through d', and point d^T must be in an arc whose radius is $c^T d'''$, and in a horizontal line through d''.

From the preceding pages, the following principles may be noted. They are also illustrated by the drawings of the previous chapters.

VIEWS OF A POINT

117. Any view of a point is a point

118. The top, front, and bottom views of a point are in the same vertical line.

119. The front, side, and back views of a point are in the same horizontal line.

VIEWS OF A STRAIGHT LINE.

120. Any view of a straight line is a line or a point.

121. Any view of a line on a plane to which it is parallel is parallel and equal to the line

122. Any view of a line on a plane to which it is perpendicular is a point

123. Any view of a line on a plane to which it is oblique is shorter than the line itself.

124. Parallel and equal lines are represented on any plane by parallel and equal lines.

125. *A straight line parallel to the front plane and perpendicular to the top plane*

The front view is a vertical line whose length equals that of the line; the top view is a point, either side view is a vertical line whose length equals that of the line.

126. *A straight line parallel to the front plane and perpendicular to the side planes.*

The front view is a horizontal line whose length equals that of the line, the top view is a horizontal line whose length equals that of the line, either side view is a point.

127. *A straight line parallel to the front plane and oblique to the top and side planes.*

The front view is an oblique line which gives the real length and the angles of the line with the top and side planes, the top view is a horizontal line shorter than the line itself, either side view is a vertical line shorter than the line itself.

128. *A straight line perpendicular to the front plane.*

The front view is a point; the top view is a vertical line whose length equals that of the line itself, either side view is a horizontal line whose length equals that of the line itself.

129. *A straight line oblique to the front plane and parallel to the top plane.*

The front view is a horizontal line shorter than the line itself; the top view is an oblique line which gives the real length and the angles of the line with the front and side planes, either side view is a horizontal line shorter than the line itself.

130. *A straight line oblique to the front plane and parallel to the side planes.*

The front view is a vertical line shorter than the line itself, the top view is a vertical line shorter than the line, either side view is an oblique line which gives the real length of the line and the angles of the line with the front and top planes

131. *A straight line oblique to the front, top, and side planes.*

The front, top, or side view is an oblique line shorter than the line itself.

Views of Plane Surfaces or Figures.

132. Any view of a plane surface or figure is a figure or a line.

133. *A plane surface or figure parallel to the front plane.*

The front view gives the real size and shape of the figure, and the real angles that its edges make with the top and side planes; the top view is a horizontal line, and either side view is a vertical line

134. *A plane surface or figure perpendicular to the front plane and parallel to the top plane.*

The front view is a horizontal line; the top view gives the real size and shape of the figure and the real angles that its edges make with the front and side planes; and either side view is a horizontal line.

135. *A plane surface or figure perpendicular to the front plane and parallel to the side planes.*

The front view is a vertical line, the top view is a vertical line; either side view gives the real size and shape of the figure and the real angles that its edges make with the front and top planes

136. *A plane surface or figure perpendicular to the front plane and oblique to the top and side planes.*

The front view is an oblique line which shows the real angles of the plane with the top and side planes; the top view is a figure which is foreshortened in one direction, either side view is a figure which is foreshortened in one direction.

137. *A plane surface or figure oblique to the front and side planes and perpendicular to the top plane.*

The front view is a figure which is foreshortened in one direction, the top view is an oblique line which shows the real angles of the plane with the front and side planes, either side view is a figure which is foreshortened in one direction.

138. *A plane surface or figure oblique to the front and top planes and perpendicular to the side planes.*

The front view is a figure which is foreshortened in one direction, the top view is a figure which is foreshortened in one direction, either side view is an oblique line which shows the real angles of the plane with the front and top planes

139. *A plane surface or figure all of whose edges are oblique to the front, top, and side planes.*

The front, top, or either side view is a figure all of whose lines are shorter than the lines of the figure they represent

140. Parallel and equal plane surfaces or figures, whose corresponding edges are parallel, are represented on any plane by similar figures whose corresponding lines are equal and parallel

LOCATIONS OF THE VIEWS.

141. The top and bottom views are always respectively above and below the front view

142. The side views are always on the same level as the front view.

143. The view of the right side is always at the right of the front view, and the view of the left side is always at the left of the front view

144. The line of any view which is nearest the front view represents the front face or front line of the object.

PROJECTION PROBLEMS.

These problems are arranged for those who wish to work by projection methods. They are not intended for public school pupils

All polygons and solids referred to are regular, unless otherwise stated. When two views are asked for, the top and front views are desired.

1. Two views of a line 2″ long. It is parallel to the top plane and 1″ below it, and at 45° to the front plane, with its nearest end ¼″ behind the front plane.

2. Two views of a line 3″ long. It is parallel to the front plane, and ⅞″ behind it, and at 60° to the top plane, from which its upper end is ½″ distant.

3 Two views of a square card whose edges are 2″ long. It is parallel to the front plane and has two edges vertical. It is ¾″ behind the front plane, and its upper edge is ¼″ below the top plane.

4. Two views of a circular card 2" in diameter. It is parallel to the front plane, with its centre 1" behind the front plane and 1¼" below the top plane

5 Two views of the same card, when it is parallel to the top plane, its centre being 1½" behind the front plane and ½" below the top plane.

6 Two views of an equilateral triangular card whose edges are 2" long. It is parallel to the front plane; its lowest edge is horizontal and 2½" below the top plane and ¾" behind the front plane.

7. (*a*) Two views of an hexagonal card whose edges are 1" long It is parallel to the front plane and ½" behind it, with two edges horizontal and the upper one ¾" below the top plane

(*b*) Two projections of the same card in the same position with reference to the top plane, but at 45° to the front plane, the nearest point being ¼" behind the front plane.

8 Two views of a prism 4" long, whose bases are squares of 2" The long edges are vertical. The upper base is ½" below the top plane, and the nearest lateral face is parallel to and 1" behind the front plane.

9. Two views of a cone whose axis is 4" long and whose base is 2" in diameter Its axis is vertical and 1½" behind the front plane, the vertex is ¼" below the top plane.

10. Two views of a pyramid whose axis is 4" long, and whose base is a square of 2" The axis is vertical and 3" behind the front plane, its upper end is 1" below the top plane. The edges of the base are at 45° to the front plane

11. Two views of an hexagonal prism 4" long. The prism is vertical, its nearest lateral face being parallel to and ½" behind the front plane. The edges of its bases are 1" long, and its upper base is 1¼" below the top plane

12 Two views of an hexagonal pyramid of the same dimensions as the prism of Problem 11, the centre of the base being 5½" below the top plane and 2¼" behind the front plane A long diagonal of the base is parallel to the front plane

13. Two views of a triangular prism 4" long, whose bases are equilateral triangles of 2" sides, when the lateral edges are horizontal and parallel to the front plane One lateral face is horizontal and

3″ below the top plane. The prism extends above this face, whose nearest edge is ¾″ behind the front plane.

14. (*a*) Two views of a prism 4″ long, whose bases are squares of 2″ side, when its lateral edges are parallel to the front plane and at 60° to the top plane. Two lateral faces are parallel to the front plane , the nearer one is 3″ behind the front plane, and its upper corner is ¼″ below the top plane

(*b*) The same prism, the same distance below and at the same angle with the top plane , but its vertical faces at 30° to the front plane. The nearest corner of the object is ½″ behind the front plane.

15. (*a*) Two views of a pyramid 4″ long, whose base is a square of 2″. The centre of the base is 2¼″ behind the front plane and 6½″ below the top plane, and the axis of the pyramid extends to the right parallel to the front plane and upward at 45° to the top plane. A vertical plane through the axis contains two of the lateral edges of the object

(*b*) Project the pyramid upon the right side plane, which is 1″ to the right of the vertex of the pyramid

16 Two views of a cylinder 4″ long and 2″ in diameter, whose axis is parallel to the front plane, at 45° to the top plane and 1¾″ behind the front plane. The upper end of the axis is 2¼″ below the top plane.

17 (*a*) Two views of a cone whose axis is 4″ long, and whose base is 2″ in diameter The lowest element of the cone is horizontal and parallel to the front plane ; it is 2½″ behind the front plane and 3½″ below the top plane

(*b*) Two views of the same cone at the same level, when the end of the horizontal element at the base of the cone is 1″ behind the front plane, the element extending from this point to the right from the front plane at 45° with it

18. The front view of a line is 2″ long and is at 60° to the front axis of projection. The top view is at 45° to this axis Give the real length of the line and measure the actual angles made by the line with the two planes.

CHAPTER IX.

SECTIONS.

145. Suppose a plane to pass through an object in any direction, the part of the object in front of the plane to be removed, and the projection of the part of the object behind the cutting plane to be made on a parallel plane. Such a projection is generally called a *section*, though section sometimes means a drawing which gives simply the actual shape of the section given by the cutting plane.

Sections are taken to show the interior of hollow objects, or the shape of solid parts which are not clearly represented by views of the outside of the object. Sections are most necessary and most used in practical working drawings but the principles involved can best be studied by the use of simple geometric solids.

146. Sections may be drawn in place of the principal views, or they may be taken in any direction whatever The position of a section should be indicated by a dot-and-dash line in the view in which the plane of the section is seen edgewise, but if the view represents the object when the part in front of the cutting plane is removed, the line of the section should be represented by a full line upon the object, and by a dot-and-dash line outside.

In practical working drawings, the objects are often shown entire in one view and in section in another In the study of projection, or working drawings, by the use of the geometric solids, all views should agree in representing the object with the part in front of the cutting plane removed In this case the part removed may be shown, if desired, by dotted or by dot-and-dash lines

147. If it is desired to represent only the line of intersection given by passing any plane through any solid, all views should agree in representing the entire solid with the line of intersection upon its surface.

148. When the part in front of the cutting plane is removed, it is customary to show the surface cut by the plane, by hatching it with parallel equidistant lines as explained in Art 84.

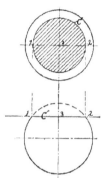

149. Sections of the Sphere. — The simplest sections are those of the sphere, for every section made by a plane is a circle, whose diameter ranges from that of the great circle, given by a plane passing through the centre of the sphere, to that of a circle as small as can be imagined, given by a plane barely cutting the sphere.

The centre *3* of the circle *C*, given by a plane which cuts the sphere and does not pass through its centre, is in a line passing through the centre of the sphere perpendicular to the cutting plane.

150. Sections of the Cube. — Sections of a cube are polygons, for the surfaces of the cube are plane, and when intersected by a plane must produce straight lines.

If a plane cuts three faces of the cube, the section is a triangle ; if it cuts four faces the section has four sides. The section may be a polygon of three, four, five, or six sides, according to the number of faces which are cut by the plane.

151. To obtain the views of the section given by any cutting plane, the plane must first be represented in the view in which it appears a line, for only in this view are the points in which it intersects the edges, or elements, of the object seen. The points of intersection, having been determined in this view, are readily projected to the other views.

152. Suppose a cube, placed so that two vertical faces are at 45° to the front plane, to be intersected by a cutting plane at 30° to the top plane and perpendicular to the front plane.

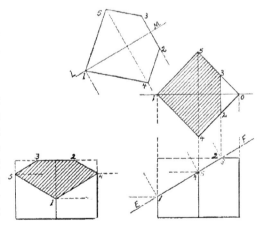

The top view of the cube is a square whose sides are at 45°; the front and side views are equal rectangles. Line *EF* in the front view represents the cutting plane. We will suppose the part of the cube above this plane to be removed, and will represent it by dotted lines. In the top view, the line *2–3* on the top face of the cube is seen of its true length and is the only edge of the section which is not represented by the square. The side view of *2–3* will be obtained by setting off one half the distance *2–3* of the top view each side of the centre of the top line of the side view, and the other points in the section will be obtained by projecting from the points, in the front view, where the plane cuts the vertical edges.

The cutting plane is oblique to the top plane and also to the side plane. The section appears of different shapes in these two views, and neither gives the real dimensions of the figure. Line *1–0* of the top view is the centre line of the section and is parallel to the front plane; therefore the front view gives the real length of the section and the distance of line *2–3* from a line connecting *4* and *5*. The top view gives the real lengths of lines *2–3* and *4–5*, and by combining the lengths of the front view with the widths of the top view, the real shape of the section will be obtained.

A simple way to obtain the true shape is to draw *LM* parallel to *EF*, and from points *1*, *2*, and *4* in *EF*, draw perpendiculars to *EF*; these perpendiculars intersecting *LM* give the lengths; the widths are obtained by setting off each side of *LM* the distances of the points in the top view from *1–0*. Join *1–4*, *4–2*, *2–3*, *3–5*, and *5–1*, and the real shape of the section is obtained.

153. The true shape of the section can be obtained as just explained only when line *1–0* of the top view is parallel to the

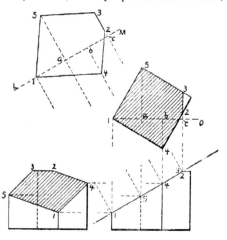

front plane. When the sides of the cube are not at 45° to the front plane the section will not be symmetrical, the front view will not give the real length of the section, and line *1–0* will pass through only one of the corners of the top view. The real shape of the section in this case can be obtained by measuring, in the top view, the distances of points *4, 2, 3,* and *5* from *1–0*, by means of perpendiculars to *1–0*, and setting these distances off on the proper lines, measuring from *LM*, which as before is drawn parallel to *1–2* of the front view and represents line *1–0* of the top view.

154. The opposite sides of the section are parallel, and whenever a plane intersects an object whose opposite surfaces are parallel, the opposite sides of the section must be parallel ; they will be of equal length when the intersected surfaces are equal in width and the intersecting plane cuts the entire width of the surfaces.

155. Sections of the Cylinder. A section of a cylinder is a circle when the cutting plane is perpendicular to the axis of the cylinder, a rectangle when the plane is parallel to the axis of the cylinder, and an ellipse when it is oblique to the axis of the cylinder.

Sections *A* and *B* require no further explanations.

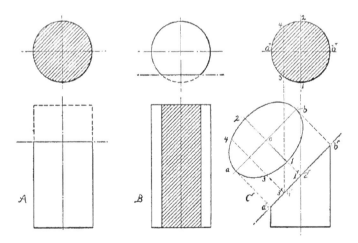

156. Section *C* is an ellipse which appears a straight line in the front, and a circle in the top view ; its real length is seen in the front view The short axis of the ellipse bisects its long axis, and intersects the axis of the cylinder; it appears a point (1^F 2^F) in the front view, and the line $1-2$ in the top view

To obtain the real shape of the section, draw $a\,b$ parallel to $a^F\,b^F$ and set off upon it the real length of the ellipse by means of perpendiculars to $a^F\,b^F$ from the points a^k, b^F, then set off each side of o, on the perpendicular to ab, one half of $1-2$ of the top view This gives the short axis of the ellipse.

To obtain other points, 3 and 4. assume a point (3^F 4^F) in $a^F\,b^F$ This represents point 3 at the front and point 4 at the back of the section. Project from 3^F 4^F to the top view and also to the real shape of the section, where the distance $3-4$ is to be made equal to $3-4$ of the top view. In this way any number of points in the real shape may be obtained.

157. If the cylinder is placed so that it does not appear a circle in any view, its section by a plane may be obtained by assuming elements upon the cylinder, and finding the points in which they are intersected by the cutting plane These points in the elements are seen in the view in which the cutting plane appears a line, and can be projected from this view to the other views.

158. Sections of a Pyramid. A section of a pyramid is a figure similar to the base, when the cutting plane is parallel to the base, a triangle, when the plane passes through the vertex and the base of the pyramid, or when it intersects the base and two of the triangular faces of the pyramid When the pyramid is intersected in any other way the section will be a polygon having a side upon each face of the pyramid cut by the plane

If the preceding articles are understood, these statements and drawings A and B will be clear to all

159. In C, plane AB cuts the lateral edges of the pyramid in four points, 1, 2, 3, 4, which are determined in the front view. Points 1 and 2 may be projected from this view to the top view. Points 3 and 4 are in lateral edges which are represented by vertical lines in both views, and therefore these points cannot be obtained in the top view by projecting from the front view

To obtain 3 and 4 in the top view of C, it is necessary to take an auxiliary cutting plane through 3 and 4 A horizontal plane through

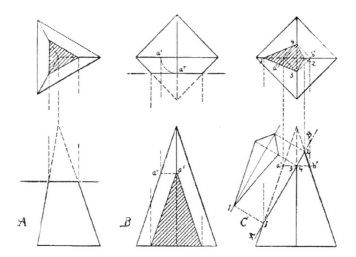

these points gives a square section whose opposite corners a^F and b^F can be projected to the top view, thus giving a^T and b^T. The corners 3 and 4 of this square are the points required to enable the top view of the section to be completed.

160. Sections of the Cone. Any section of a right circular cone made by a plane parallel to its base is a circle. This section is illustrated by Fig. 88.

161. If a cutting plane passes through the vertex of the cone, it intersects the base of the cone in a chord of the circle, and the curved surface of the cone in two elements; the section is thus a triangle. See Fig. 96.

162. If the plane intersects all the elements of the cone, the section is an ellipse. Fig. 98.

163. If a cone is intersected by a plane which is parallel to one of the elements of the cone, the section is a parabola. Fig. 99.

164. If a cone is intersected by a plane which makes a greater angle with the base than do the elements, the section is an hyperbola. Fig. 97.

165. When a solid, bounded by plane surfaces, is cut by any plane, the edges of the solid are seen piercing the plane, in the view in which the cutting plane appears a line; these points of intersection are readily projected to the lines of the other views, and thus the angles of the section are obtained.

166. When curved bodies or those with curved surfaces are intersected by a cutting plane, only the points of intersection in the contour elements are seen in the view in which the cutting plane appears a line. To obtain other points in the section, it is necessary to assume elements or other lines upon the curved surface, their intersections will be found in the same way as the intersections of the edges of solids bounded by plane surfaces. When possible the lines assumed should be elements of the surface. Fig. 98.

Instead of assuming elements in order to obtain the points in a section, it is often easier and more accurate to pass through the object a number of auxiliary cutting planes whose sections are simple. These planes intersect the plane of the section in lines, whose ends must be points in the required section. See *4* and *5*, Fig. 97.

If pupils can obtain the sections of the regular geometric solids they will readily find the sections of any solid which may be constructed.

CHAPTER X.

INTERSECTIONS.

167. In practical work it is necessary to represent all kinds of regular and irregular bodies which intersect or penetrate each other. The knowledge required to do this is best obtained by study of the geometric solids

Simple intersections are produced when a body small enough to pass through another enters one plane surface and leaves by another. The large object being bounded by plane surfaces or at least by the two mentioned, the intersections are simply sections of the smaller body made by the plane surfaces of the larger, and have been explained

Simple intersections are also given by a cone or cylinder which penetrates a sphere in such a way that its axis passes through the centre of the sphere In this case the plane of the intersection must be at right angles to the axis of the penetrating body and the lines in which the cylinder or cone enters and leaves the sphere must be circles Fig. 100.

If the axis of the cone or cylinder does not pass through the centre of the sphere, the intersections will not be circles, and must be obtained as explained later.

168. The principles of sections enable us to find all intersections, for if the lines of intersection are not, in part or in whole, some of the regular sections explained, they can be determined by means of points situated in sections given by auxiliary cutting planes

169. When bodies bounded by plane surfaces intersect, the lines of intersection will be straight, and must connect the points in which the edges of each solid penetrate the other solid . the problem is thus to find these intersections

170. When curved bodies intersect or are intersected. there are elements, instead of edges, which penetrate the surfaces and must be treated as if they were edges. Thus problems in intersections, as they are generally solved, may be reduced to the simple problem of finding the intersection of a line and a plane, or of a line and a curved surface.

INTERSECTIONS OF A LINE AND A PLANE SURFACE.

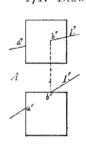

171. Drawing A represents a cube pierced by an inclined line which enters the left side at a and leaves the top of the cube at b. The front view determines both a and b, for it represents both the left and the top surfaces by straight lines. The top view determines only one point (a), for in it the top of the cube appears a surface, and *when a plane appears a surface its intersection by a line cannot be determined by means of this view alone.* Point b must be obtained in the top view by projecting from the front view.

NOTE. — The positions of the intersecting lines in all the problems are assumed.

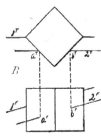

172. Drawing B represents a cube and a line 1 which is in the same plane as the right and left vertical edges of the cube, and therefore intersects these edges. Line 2 is in front of the right and left vertical edges; its intersections with the faces of the cube are seen in the top view, and may be projected from this view to the front view.

173. Drawing C represents a cube intersected by a line which

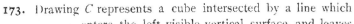

enters the left visible vertical surface, and leaves at the top of the cube. The front view represents the top of the cube by a horizontal line, and in this view the intersection b^F of the line with the top is seen.

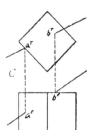

The intersection a of the line with the left vertical surface of the cube is seen in the top view. The points determined by each view can be projected to the other view, and in this way each view completes the other

174. Drawing D shows a line which pierces the left end and the

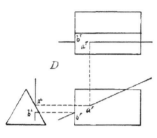

front inclined surface of a triangular prism. The intersection of the line and the inclined surface cannot be determined in either the front or the top view, for in neither view does the surface appear a line. In the side view the inclined surface appears a line, and the intersection a^L is seen and may be projected to the other views.

175. Drawing E represents a square pyramid and an inclined

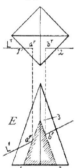

line L which pierces the pyramid in front of the axis. The intersections of the line and the lateral faces of the pyramid cannot be seen in either view, for the faces appear surfaces in both; and for the same reason a side view is of no assistance. To find the points, an auxiliary cutting plane must be taken through the line. If a vertical plane is chosen it will appear the line L^T, in the top view, and in this view its intersections, *1* and *2*, with the edges of the base are seen. It intersects the central lateral edge in *3*, which is found as explained in Art. 159. In the front view the triangle *1, 2, 3* is the section made upon the pyramid by the vertical cutting plane passing through L. The points a and b, where L intersects the sides of the triangle, are the points in which L enters and leaves the surface of the pyramid.

176. Drawing F represents the same pyramid and line as draw-

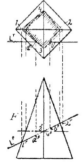

ing E, but instead of choosing a vertical plane through L as the auxiliary plane, a plane is passed through L, perpendicular to the front plane. The front view gives the points 1^F, 2^F, 3^F, and 4^F where the lateral edges intersect this cutting plane, whose figure of intersection is readily found as explained in Art. 159. In the top view this section appears a surface, and the points in it, a^T and b^T, in which L enters and leaves the pyramid, are seen and may be projected to the front view.

177. Drawing G represents the same conditions as D.

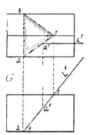

The intersection may be found without drawing the end view, by means of a cutting plane used as explained in E and F.

The cutting plane used in the illustration is perpendicular to the front plane; and in the front view points 1, 2, 3 of the section, which are in the edges of the prism, are seen. When these points are projected to the top view and connected, the triangular section is represented by a triangle, and the point a^T, where L^T intersects the triangle, is the top view of the required point a.

INTERSECTIONS OF A LINE AND A CURVED SURFACE.

178. Drawing H represents a sphere intersected by a horizontal line L, in front of the centre of the sphere.

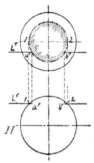

If the line were in a vertical plane passing through the centre of the sphere and parallel to the front plane, the front view would give the intersections of the line and the sphere. If the line were in a horizontal plane passing through the centre of the sphere, the top view would give the intersections of the line and the sphere. As the line is not situated in either of these positions, the intersections cannot be in the circle of either view, and must be found by means of an auxiliary cutting plane.

To obtain the points in which L enters and leaves the sphere, a horizontal cutting plane may be passed through the line. This plane gives a circular section whose diameter 1–2 is seen in the front view. The circle appears a circle in the top view; here its intersections a^T and b^T with L^T are seen, and may be projected to the front view.

179. Drawing I solves the same problem, except that the line is parallel to the front plane only. An auxiliary cutting plane parallel to the front plane is used. This plane gives a circle on the sphere whose diameter, *1–2*, is seen in the top view, its real shape is seen

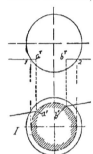

in the front view, where *a* and *b* are determined, and from which they may be projected to the top view.

When *L* is parallel to only one plane, the section must be taken through *L* parallel to this plane. If *L* is not parallel to either plane the cutting plane may be perpendicular to either plane and will give a circle which in one view will appear an ellipse.

180. Drawing *J* represents a sphere and two inclined lines which are represented in the top

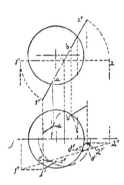

view by the line *1^T–2^T*, and in the front view by two lines, the upper of which intersects the sphere in points *a* and *b* and has its ends unmarked; the lower line intersects the sphere in points *c* and *d*, and its ends are represented by points *1^F* and *2^F*. The upper line *1–2* intersects the sphere at points *a* and *b*, which are upon the upper surface of the sphere, point *a* being upon the front surface. Neither *a* nor *b* appears, in either view, in the contour of the sphere, and to determine these points, a vertical cutting plane is taken through the

line. This plane intersects the sphere in a circle, which in the front view appears an ellipse. Only the upper half of this ellipse is drawn, as it contains both *a* and *b*. (When following this solution, cover the lower half of the front view with paper, to hide the second solution of the problem.)

181. If the lower half of the ellipse, which represents the section given by the vertical plane through the lines, is drawn, it will contain points *c* and *d*, in which the lower line intersects the surface. These points may also be found by revolving the vertical plane taken through *1–2*, until it is parallel to the front plane. The section given by it will then appear a circle in the front view. As the plane revolves about a vertical axis passing through the centre of the circular section, the points of the section and points *1* and *2* describe arcs of circles, which appear arcs in the top view, and horizontal lines in the front view. (The upper half of the front view is to be covered when following this description.)

As the vertical cutting plane passing through line *1–2* revolves into a position parallel to the front plane, points *1* and *2* move, in the front view, in horizontal lines. In these lines points *1″* and *2″* must be given by projecting lines from *1′* and *2′* which represent the ends of the line *1–2* when it is parallel to the front plane. When the vertical plane passed through *1–2* has been revolved parallel to the front plane, the circle which it gives upon the sphere will appear a circle, whose lower half is shown in the drawing, the line *1″–2″* intersects this circle in points *c′* and *d′* These points move, when the plane of the section is revolved back into its original position, in arcs which appear horizontal lines in the front view, and which intersect *1^F–2^F* in *c^F* and *d^F*, these are the front views of points *c* and *d*, in which the lower line *1–2* intersects the surface of the sphere. These points are not shown in the top view, but would be given by verticals from *c^F* and *d^F*, intersecting *1^T–2^T*

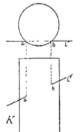

182. Drawing *K* represents a cylinder intersected by a line *L*

The intersection of a line and the curved surface of a cylinder is determined in the view in which the curved surface appears a circle If the curved surface is not seen edgewise in any view, the cylinder must be cut by a plane passing through the line, in order that the points of intersection may be determined, this applies to any and all positions of the cylinder and the line.

183. Drawing *L* represents a cone, and a line which intersects it in front of its axis.

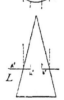

If the line were in the plane of the contour elements, the points of intersection would be seen in the front view, but, as this is not the case, to obtain the points, an auxiliary plane must be used A horizontal cutting plane, *A*, intersects the cone in a horizontal circle, which appears a circle in the top view In this view the intersections, *a* and *b*, of the line *A* and the circle are determined, and may be projected to the front view

184. Drawing *M* represents a cone intersected by an inclined line.

To obtain the points of intersection, a cutting plane perpendicular to the front plane may be used; this will cut the cone in an ellipse, which will appear an ellipse in the top view. A cutting plane perpendicular to the top plane is, however, used here. This cuts the cone in an hyperbola, whose real shape is seen in the front view. Art. 164. The intersections, *a* and *b*, of *L* and the cone are seen in the front view, and from this view may be projected to the top view.

185. When auxiliary cutting planes are used, it is not necessary to find the complete section given by the plane, as the part near the point of intersection is all that is required.

INTERSECTIONS OF SOLIDS.

186. All the points necessarily involved in intersections are explained above. If understood they can be readily applied when, instead of a single line penetrating a solid, one solid penetrates another. In this case, when the points of intersection of the various lines of the penetrating solid cannot be seen in one view or the other, it is necessary to use auxiliary cutting planes. These planes should be so chosen as to give simple sections upon both solids. The circle and the rectangle are the simple sections of the cylinder ; the triangle and the circle are the simple sections of the cone ; and in many problems it will be possible to obtain these sections instead of ellipses, hyperbolas, and parabolas.

187. When a cutting plane intersects two intersecting bodies, it gives upon each a certain figure. These figures intersect in points, which must be points in the lines in which the solids intersect.

This is illustrated by *N* and *O*. In *N*, a vertical cutting plane intersects the horizontal prism in a rectangle, and the sphere in a circle. These figures intersect in four points, which are points in the lines of intersection of the prism and the sphere.

In *O*, a horizontal cutting plane gives a rectangle upon the cylin-

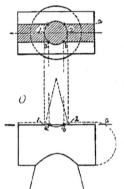

der and a circle upon the cone. These figures intersect in four points, which are points
in the lines of intersection of the cylinder and
the cone.

The same problem in intersections may be
solved in many different ways, but always by
means of auxiliary cutting planes, when the
required points are not seen at once in one of
the given views. These planes may be taken
in so many different positions that to explain
all would be impossible in the limits of this
book. If the principles of the subject are
understood, pupils will have no trouble in
deciding what planes will give the simplest sections, or in determining the points of these sections.

Instead of using parallel cutting planes, as explained in the preceding articles, one plane may be supposed to be hinged, or to swing
upon a given line as an axis. In the case of two intersecting pyramids or cones, this axis should pass through the vertices of both
solids, thus the sections of both will be triangles. If one solid is a
cylinder or prism the plane should swing upon a line parallel to the
axis of this solid. This method is not illustrated, but is often of
great value.

CHAPTER XI.

ARRANGEMENT AND NAMES OF VIEWS.

188. In descriptive geometry the planes of projection are supposed to be indefinite in extent, one horizontal and the other vertical. See Fig. 35. Four dihedral angles are formed by these planes. The eye is supposed to be always in front of the vertical plane and above the horizontal plane. Objects may be placed in any of these angles and projected upon the two planes; in projection and descriptive geometry all four angles are used.

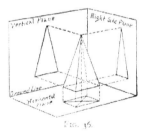

Fig. 35.

The projection of an object upon the front vertical plane is called its vertical projection; that upon the horizontal plane is called its horizontal projection. These drawings are also called elevation and plan, and front view and top view. All these terms are applied in practical work, but the preference seems to be to call shop drawings views. In the study of projection and descriptive geometry, the drawings are generally called projections.

189. If an object is in the first angle, it is between the planes and the eye, and covers its projections upon the planes.

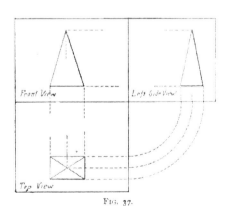

Fig. 36.

Fig. 37.

Fig. 36 represents the vertical and horizontal planes forming the first angle, with a side vertical plane placed at the right. The drawing also shows a pyramid placed within this angle, and its projections upon the three planes.

When the planes are revolved, as in Fig. 37, the top view of the pyramid is found below the front view, and the view of the left side comes at the right of the front view.

This arrangement is very different from that due to the use of the third angle, which is explained in Chap. III, and in which the top view is above the front view.

190. Draughtsmen have no uniform way of working. Some arrange the views according to the use of the first angle, and some according to that of the third angle. To those who do not read drawings easily this lack of uniformity results in confusion. To those who are accustomed to reading them, the positions of the views are of less importance, and the convenience of the draughtsman in making the drawings should determine largely the method of arrangement.

It may often happen that it is not easy or convenient to adhere to any system, and it is certainly reasonable that a rule which hampers should not be followed. Generally, however, some uniform arrangement of views may easily be given, and most draughtsmen use either that given by the first or that given by the third angle. The chief points to decide the arrangement to be adopted should be the ease and accuracy of making and of reading the drawing.

191. If an object similar to that illustrated by Fig. 38 be placed in the first angle, the draughtsman will be obliged to project from the right end of the front view across its entire length to the space at the left of the front view, to make the view of the right end of the object, and in the same way across the entire length of the front view to give the view of the left end. He will project from the top of the front view to the space below this view to draw the top view; and from the bottom of the front view to the space above this view to draw the bottom view. To make drawings thus arranged must take much longer than to make them arranged as in Fig. 38, with the view at the left of the front view showing the left end of the object, that at the right, the right end, that below, the bottom of the object, and

so on. Not only will it take longer to make the drawings, but the inaccuracies of drawing boards and T-squares will cause the drawings to be less exact than those whose arrangement is due to the use of the planes placed in front of the object.

The views due to the use of the first angle are not so easy to read as those due to the use of the third, for the third angle places the different views of the same part as near as possible to each other, while the first angle places them as far apart as possible. These considerations make the use of the third angle desirable for all prac-

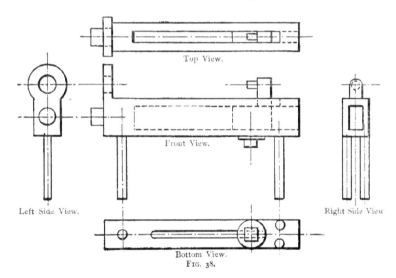

Top View.

Front View.

Left Side View.   Right Side View

Bottom View.
FIG. 38.

tical working drawings; and therefore, this arrangement is adopted throughout this book.

  192. In the study of projection with advanced classes the choice of angle is of little importance. The objects may be placed in either of the angles, and there will be little difference in the ease (or perhaps, difficulty) with which the students understand the subject. A point in favor of the first angle for use in the study of projection is that books on the subject of projection generally use this angle; another point is that when the third angle is used, and glass planes and expensive models to place behind them are not provided, the problems cannot be easily illustrated. When the first angle is used two blackboards may be hinged together, as illustrated in Fig 36 ; they should be arranged so that they may revolve while fixed at right angles to each other, or revolve independently so as to form one vertical surface. Thus the students may see the horizontal board edgewise and the vertical one as a surface ; or the horizontal one as a surface and the vertical one edgewise ; or with the plane of the horizontal one coinciding with that of the vertical, as in Fig 37. A vertical blackboard may be attached at either side of the vertical board, as illustrated in Fig. 36, to represent a side vertical plane. By means of these boards and of models which can be procured with little expense and held as desired in front of the planes, the problems may be much more easily illustrated than when the third angle is used. These boards are readily managed and can be seen by all the occupants of a large class-room, so that for advanced classes who are studying the theory of projection, the first angle seems preferable, especially when the course is preparatory to one in descriptive geometry. But although books on this subject use principally the first angle, all are necessary, and a thorough course will deal with points, lines, planes, and objects in all the angles.

The third angle as well as the first might be used principally in the study of descriptive geometry. The fact that the arrangement of views given by the first angle is different from that which we have decided to be the best for shop drawings should not, however, be of the slightest consequence to the advanced student, for he should study the relations of points in all the angles ; and he should be as familiar with the arrangement due to the third angle as he is with that due to the first. When his course is completed and he makes practical drawings, he will then make them with equal ease, whatever arrangement of views he may choose

Projection and descriptive geometry are means by which the mind is trained to conceive any form in any position ; their aim is not the simpler and less important one of enabling young students to make views or working drawings of concrete objects.

The question as to whether to use solids, planes, lines, or points for the first study, has caused some discussion From what has been said, it will be seen that for the first work in the public schools and for elementary work generally, it will be better to begin with solids , for to deal with lines, etc., is more in accordance with projection than with the making of views from simple objects.

In the theoretical study of projection it makes no difference what is taken first; for if the solid is taken the lesson still must deal with the relations of points and lines to each other and to the planes of projection, so that in reality these are studied first.

CHAPTER XII.

PLATES AND EXPLANATIONS.

The plates of this chapter illustrate work from that suited for the youngest pupils of the subject, to that which advanced pupils of high and elementary technical schools may require. Teachers are to select work suitable for their pupils and make it as practical as possible by following the directions given in the preceding chapters.

The principles of working drawings may be taught by means of free-hand sketches. Free-hand sketches may also be made and dimensioned in order that finished drawings may be made from them All other drawings should be made by the use of instruments, and to some fixed scale In the lower grades of the public schools an architect's scale cannot be provided, and, when the objects studied cannot be represented full or half size, a special scale must be drawn as explained in Art. 10.

The drawings of the plates are small, and, in order that they may not be obscured, working or projecting lines are given only when necessary to illustrate important principles or ways of working; and dimensions are given only in a few cases, but sufficient to show clearly how dimensions should be placed. In some of the drawings, dimension lines and arrow-heads are placed where dimensions should be given this is done so that the drawings may not be used for copies, and that objects similar to those illustrated may be studied, measured, and drawn to scale

The models represented are often different in proportion from the regular drawing models , the proportions are chosen to present the principles by drawings as large as the plates will allow.

Those who understand the preceding chapters will require no further explanation of the points there considered; therefore explanations of such points, if given at all, will be stated very briefly, and often simply by reference to preceding articles.

The plates are reproduced from drawings twice their size. The lines of the original drawings are suitable in width for practical drawings. The lines of the plates are finer than is required for any except the most highly finished drawings made by advanced pupils

The lines of the drawings upon pages 61 and 106 are suitable in width for practical drawings and for pupils' work.

PLATE I.

FIG. 39. *Front and top views of a sphere.*

Any view of a sphere must be a circle, whose diameter is equal to that of the sphere. The centres of these views are in a vertical line.

FIG. 40 *Front and top views of an hemisphere whose plane surface is horizontal and uppermost.*

The horizontal circle is represented by a circle in the top view, and by a horizontal line in the front view (Art 134), the curved surface of the hemisphere is represented in the front view by a semi-circle extending from the horizontal line downward.

FIG. 41. *Front and top views of a cube, two faces being horizontal, and one face appearing a square in the front view.*

The top face appears a square in the top view.

For the development and that of any simple solid in the following figures, see Chap. IV.

FIG. 42. *Front and top views and development of a right square prism*

FIG. 43 *Front, top, and right side views of a vertical square tablet so placed that the front view gives its real shape.*

See Art. 133.

FIG. 44. *Front and top views of a vertical oblong tablet, which appears its real shape in the front view.*

See Art. 133.

FIG 45 *Front and left side views, and development of a horizontal cylinder, whose axis is parallel to the front plane.*

The front view is a rectangle and represents the circles by vertical lines; the distance between these verticals is equal to the length of the cylinder. The side view is a circle, and should be drawn first.

The form of the laps by which the parts are fastened together is immaterial : that shown in the figure may, in the case of the cylinder or cone, give the most satisfactory results.

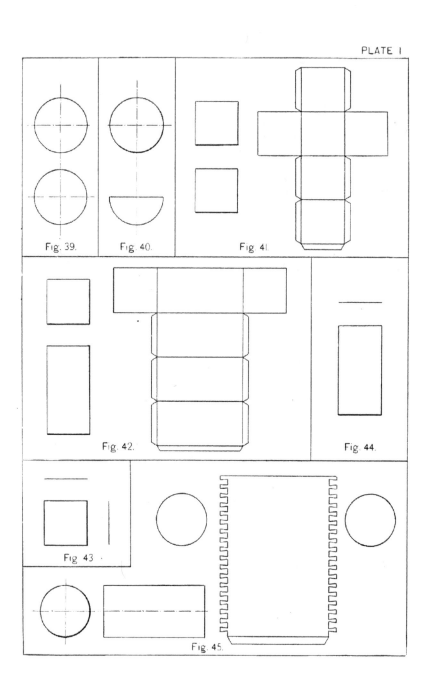

PLATE I

Fig. 39.

Fig. 40.

Fig. 41.

Fig. 42.

Fig. 43.

Fig. 44.

Fig. 45.

PLATE II.

FIG 46   *Front and left side views, and development of an equilateral triangular prism, placed horizontally and so that the side view is a triangle*

The side view gives the real shape of the triangle and should be drawn first (Art 135).

The front view is an oblong, whose length is equal to that of the prism, and whose height, given by projecting from the side view, is equal to *1–2* of that view (Art 138).

FIG 47   *Front and top views of a circular tablet which appears its real shape in the front view.*

See Art 133

FIG 48.   *Front and top views of an equilateral triangular tablet which appears its real shape in the front view*

See Art 133

FIG. 49.   *Front and top views and development of an upright square pyramid, placed so that the edges of its base are at 45° to the front plane.*

The top view is a square and should be drawn first (Art 134)   The lateral edges are represented in this view by the diagonals of the square In the front view the length of the axis is seen, also the real length of the right and left lateral edges , the nearest lateral edge appears a vertical line as long as the axis, and thus shorter than its actual length (Art 130).

FIG. 50.   *Front and top views of a vertical tumbler.*

At the top and bottom of the object are horizontal circles, which appear concentric in the top view. Draw the top view of the outside upper edge of the tumbler ; then the front view of the tumbler ; and from the dotted line, showing the thickness of the glass, obtain the dimensions of the inside circles of the top view.

FIG. 51   *Front and top views of a tin cookie-cutter, placed so that in the front view the handle is a semi-circle.*

The top view of the cutter is a circle and should be drawn first ; then draw the front view of the cutter and add the handle ; draw last the handle in the top view.

FIG. 52.   *Top and front views of a cylindrical box.*

The thickness of the top and bottom is shown by the dotted lines

Draw the top view first (Art 134)

FIG 53   *Front and top views of a tin dipper.*

Draw the top view of the dipper first ; then the front view , then the front view of the handle , and lastly the top view of the handle.

PLATE II.

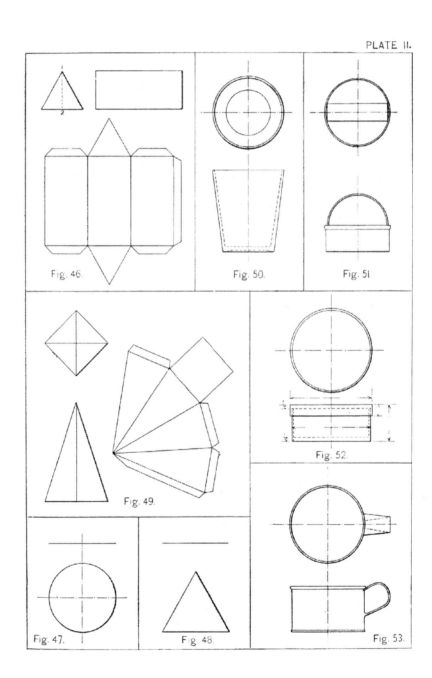

Fig. 46.

Fig. 50.

Fig. 51

Fig. 49.

Fig. 52.

Fig. 47.

Fig 48.

Fig. 53.

PLATE III.

FIG. 54. *Top and front views of an upright hexagonal prism, and the development of the same.*

FIG. 55 *Top and front views and development of an upright hexagonal pyramid.*

FIG. 56 *Front and top views and development of an upright cone.*

FIG. 57. *Front, top, and right side views of an hexagonal tablet.*

FIG. 58. *Front and top views of a tin tunnel, placed so that its circular edges appear circles in the top view.*

FIG 59. *Top and front views of a tin grater.*

In all the figures of this plate, except Fig. 57, the top view should be drawn first. In Fig. 57 the front view should be drawn first.

PLATE III.

Fig. 54.

Fig. 55.

Fig. 56.

Fig. 57.

Fig. 58.

Fig. 59.

.

PLATE IV.

Fɪɢ 60   *Front, top, and right side views of a knife-box.*

Fɪɢ 61   *Top and front views of an oil-can.*

Draw the top view first.

Fɪɢ. 62.   *Top and front views of a call-bell.*

Fɪɢ 63   *Front and left side views of a horizontal hollow cylinder; a horizontal section on AB, and a cross section of the cylinder on CD.*

Draw the side view or the cross section first.

Fɪɢ. 64.   *Top and front views of a dish with a handle.*

Draw the top view of the dish first; then the front view; then draw the handle.

Fɪɢ. 65.   *Front and top views of a mallet.*

Draw the front view of the head first, and then carry the drawing of the two views along together, as explained in Art 36

All the above drawings may be made most advantageously by carrying all the views of each object along at the same time, as explained in Art. 36 It is, however, not necessary or important that young pupils should work in this way. With them the question is not speed, but accuracy and understanding of the principles · hence they may begin with the view which is easiest to draw, and may carry it as far as possible, and even finish it, before beginning any other view.

.

PLATE IV.

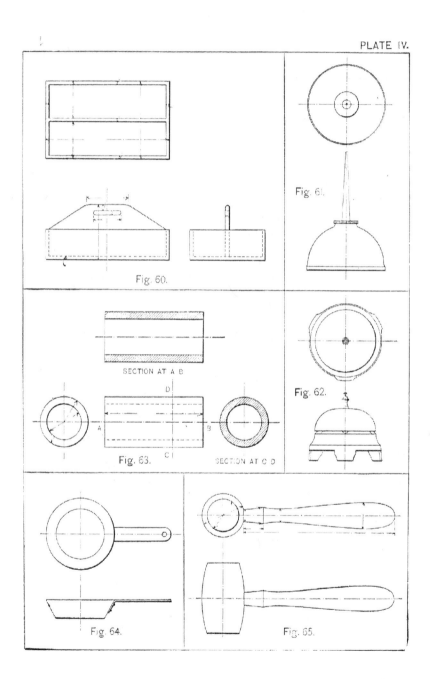

Fig. 60.

Fig. 61.

SECTION AT A B

D

A B

Fig. 63.

C

SECTION AT C D

Fig. 62.

Fig. 64.

Fig. 65.

PLATE V.

FIG. 66.   *Top and front views of a tin coffee-pot.*

Draw the top view of the pot first; then the front view, adding the handle and nose, first in this view, and then in the top view

FIG. 67.   *Front, and left side views, and a longitudinal section of a tool handle.*

Draw the side (end) view first, and the other two views as explained in Art 36.   The handle, being of wood, may be represented, if desired, as shown on page 61.

FIG. 68.   *Front, top, and left side views of a flatiron*

FIG. 69   *A front view, a section on CD, and a horizontal section at AB through a horizontal spool.*

Draw the section on *CD* first

FIG. 70.   *Front and top views of a vertical circular tablet, attached to the back edge of a horizontal square tablet*

FIG. 71   *Front and top views of a vertical hexagonal tablet, attached to the back edge of a horizontal square tablet*

PLATE V.

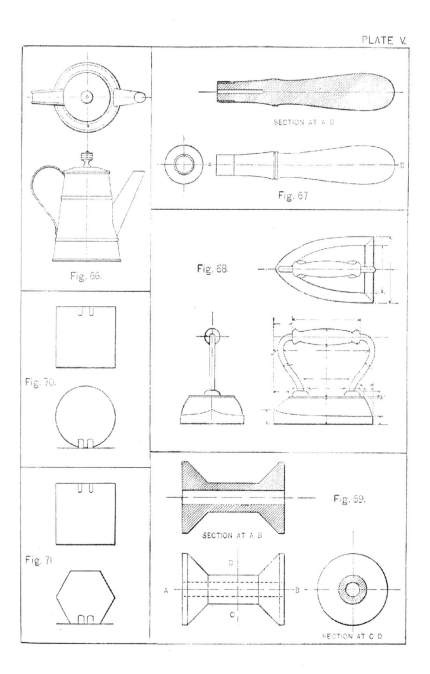

Fig. 66.

Fig. 67.

SECTION AT A B

Fig. 68.

Fig. 70.

Fig. 71

Fig. 69.

SECTION AT A B

SECTION AT C D

PLATE VI.

FIG. 72   *Top, front, and right side views of a horizontal oblong tablet, with a vertical triangular one attached to its right edge*

Draw the top or side view first

FIG. 73.   The same views of the same combination as in Fig. 72, but with a vertical pentagonal tablet attached at the back edge of the horizontal tablet.

Represent the pentagonal tablet in the front view first.

FIG 74   *Front, top, and left side views of a chair, represented by combining a horizontal square tablet for the seat with vertical square tablets for the front and back of the lower part, and a vertical square tablet for the back of the chair.*

Draw the top view first.

FIG. 75.   *Front, top, and right side views of a square prism, supporting an oblong tablet at an angle of 45°.*

Draw the side view of the prism first; then the front and top views. Draw the front view of the tablet first, then the side and top views.

FIG. 76   *Front, top, and right side views of a circular plinth, or of tablets arranged in the form of a plinth, supporting a square plinth, or tablets arranged in the form of a plinth*

Draw the top view of the circular plinth first, then the front and side views   Draw the front view of the square plinth first, and then the top and side views.

FIG 77   *Top and front views of a square plinth with a circular plinth resting upon it.*

Draw the top view first.

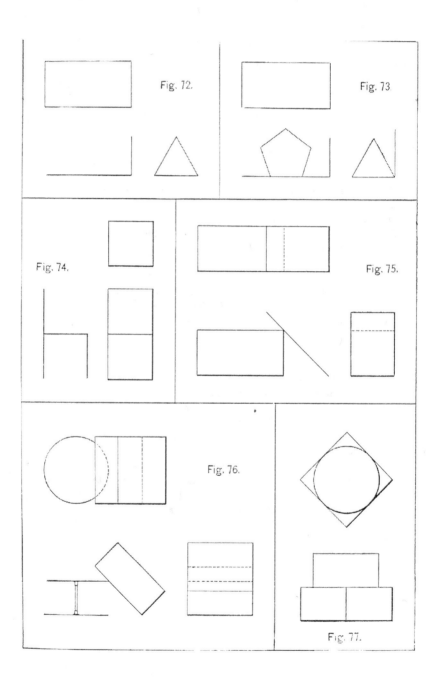

Fig. 72.

Fig. 73

Fig. 74.

Fig. 75.

Fig. 76.

Fig. 77.

PLATE VII

FIG 78　*Top, front, and right side views of an hexagonal plinth with a circular plinth upon it*

Draw the top view first　Make the width of the side view equal to *1–2* of the top view

FIG 79　*Top front, and right side views of a circular plinth with a square plinth upon it*

Draw the top view first.

FIG 80　*Front top, and right side views of a triangular prism supporting an oblong tablet at an angle of 45°*

Draw the side view first

FIG 81　(*A*) *Front and top views of an hexagonal tablet, when it is parallel to the front plane*

(*B*) *Front and top views of the same hexagonal tablet when it is at an angle of 60° to the front plane and perpendicular to the top plane, as in A*

Draw the front view of *A* first　then the top view　Place the top view at 60° and project from it and from the front view of *A*, to complete *B* as explained in Art 109

FIG 82. (*A*) *Front and top views of a square prism, two of its oblong faces being at 45° to the top plane and the other two being parallel to the front plane*

(*B*) *Front and top views of the same square prism when two of its oblong faces are at 45° with the top plane, and the other two are at 30° to the front plane.*

Draw the front view of *A* first, then the top view　For the top view of *B* place the top view of *A* at 30° and then obtain the front view of *B*, as explained in Arts 112 and 113.

At *C* are given front and top views of the prism when in an upright position

The exponents *F*, *T* and *S* used in this and following figures indicate the different views, — front, top, and side　To avoid confusion they are not placed upon all the figures and some of the points are not numbered or lettered

PLATE VII.

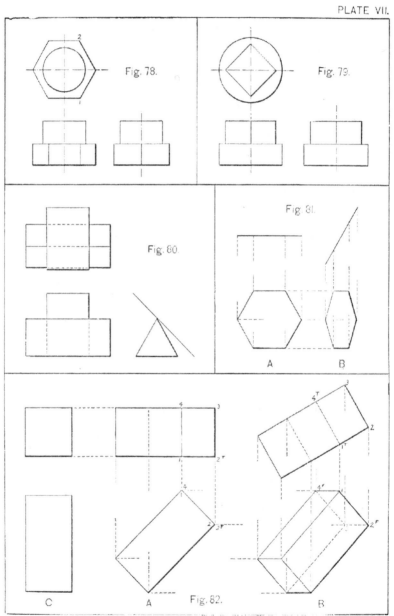

Fig. 78.

Fig. 79.

Fig. 80.

Fig. 81.

A   B

C   A   Fig. 82.   B

PLATE VIII.

FIG. 83. *Front, top, and right side views of the lower portion of an upright square prism, cut by a plane at an angle of 45° to its base, also the real shape of the section and the development of the surface of the object.*

Draw the top view first, next the front and side views, then, in the front view, the line of the cutting plane at 45°. Project the section to the side view and obtain its real shape as explained in Art. 152.

Develop the lateral surface of the prism as explained in Art 54. To obtain the line of intersection upon it, measure (in the front or side view) the distances of points *1, 2, 3,* and *4* from the base of the object and set these distances off on the development, obtaining 1^D, 2^D, 3^D, and 4^D. Joining these points in the lines defining each lateral surface gives the line of intersection. Place the bases upon any desired lines of the development of the bases.

FIG. 84. *Top, front, and right side views of an upright square prism whose lateral faces are at 45° to the front plane, and which is intersected by a plane at an angle with its base, also the real shape of the section and the development of the lateral surface of the object.*

See Arts. 152 and 54.

FIG. 85. *Front, top, and right side views of a horizontal triangular prism; a section of the same by a plane at 60° to its horizontal face; also the development of its lateral faces.*

Draw the side view first, then the front and top views, then the line of the section at 60° in the front view, then the top view of the section, and last the real shape of the section. The lengths $A-2^D$, $B-3^D$, etc., for the development are seen in the front and top views. The distance between A and B, etc., is seen in the side view.

PLATE VIII.

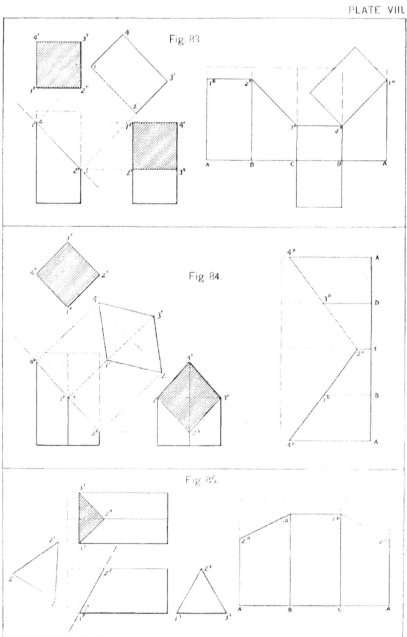

Fig. 83.

Fig. 84.

Fig. 85.

PLATE IX

FIG. 86. *Top, front, and right side views of a vertical hexagonal prism cut by a plane at 45° to its base, also the real shape of the section and the development of the surface*

FIG. 87 *Top and front views of the frustum of a square pyramid, and the development of its surface*

Draw the top view of the complete pyramid first · next its front view, and then the section in the front view, project from the front view the points for the square which is the top view of the section and shows its real shape.

The length of the lateral edges of the entire object is seen in the front view from 1^F to 2^F, and the length of the part below the cutting plane is seen from 2^F to 3^F The sides of the base appear their real length in the top view

See Art 59

FIG. 88 *Top and front views and development of the frustum of a cone*

To develop, divide one quarter of the base into three equal parts Draw an arc with radius 1^F-2^F Upon this arc set off 2^T-4^T twelve times. The development of the line of the section may be obtained by considering the section as the base of a cone extending from the vertex to the cutting plane The elements of this cone are equal in length to the distance 1^F-3^F

FIG 89. *Top, front, and right side views and development of an upright square pyramid, intersected by a plane at 45° to its base*

The lateral edges of the pyramid are not parallel to the front or side planes, and do not appear their real lengths in the front or side view To develop the surface, the length of the lateral edges must be found by revolving one of them, as 1-2, until in the top view it is horizontal and at 1^T-$2'$ As the pyramid revolves upon its axis until the lateral edge comes to this position, point 2 moves in an arc which appears an arc in the top view and a horizontal line in the front view, and $2''$ will be under $2'$, and on the level of 2^F, and the distance $2''$-1^F must be the real length of the edge 1-2. Point 3. in 1-2, also describes an arc in the top view, but it is not necessary to draw the arc, as $3''$ must be in 1^F-$2''$ and on the level of 3^F. The distance 1^F-$3''$ is the real distance from the vertex to the points of the section which are in the two left lateral edges. The real length of the distance, 1-4, from the vertex to the points in the two right lateral edges is found in a similar way

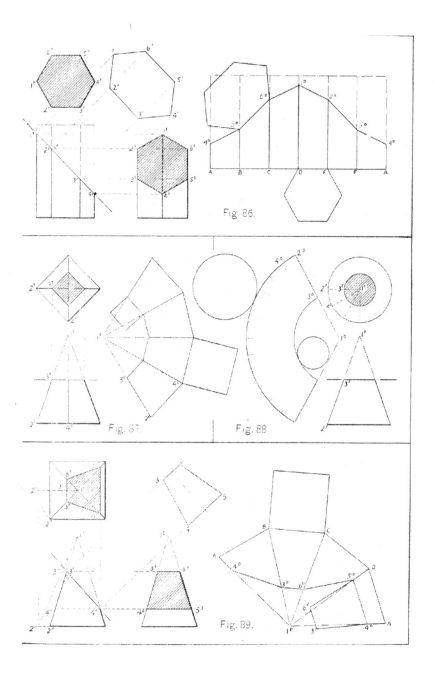

Fig. 86.

Fig. 87.

Fig. 88.

Fig. 89.

PLATE X.

FIG 90   (*A*) *Front and top views of a vertical circular card, parallel to the front plane*

(*B*) *The same views of the same card when perpendicular to the top plane and at an angle of 45° with the front plane.*

The views of the first position are necessary to obtain those of the second, and the top view of *A*, when it is placed at 45°, will be the top view of *B*.

Draw the horizontal and vertical diameters in the front view of *A* and mark their ends *1, 3* and *2, 4*   Place these points in both the top views and obtain them in the front view of *B* by verticals from the top view of *B* intersected by horizontals from the front view of *A*   These points are the ends of the diameters of the ellipse, which is the front view of *B*   To find other points, place any point as C^F in the front view of *A*. Then place C^T in both top views, and find *C* in the front view of *B* as explained. If more than four points are desired it is best to take eight or twelve equidistant points.

FIG 91   (*A*) *Front and top views of a cone resting upon an element which is horizontal and parallel to the front plane.*

(*B*) *Top and front views of the same cone, when the element upon which it rests is horizontal and at an angle of 30° to the front plane.*

The front view of *A* will be an isosceles triangle of which the lower of the two equal sides is horizontal ; the base of the triangle represents the base of the cone   If the dimensions of the base and axis of the cone are given, the length of the elements must be found by drawing an isosceles triangle whose base is equal to the diameter of the base of the cone and whose altitude is equal to the axis of the cone.

To obtain the front view of *A*, draw the element *1–9* horizontal and its real length. With 9^F as a centre and 9^F–1^F as radius, describe an arc through point 1^F. With 1^F as a centre and the diameter of the base of the cone as a radius, describe an arc to intersect the arc from 9^F in point 5^F. Join 1^F–5^F and 5^F–9^F, this gives the front view of the cone   Bisect the line 1^F–5^F and obtain the centre of the base of the cone

The top view of the base is an ellipse which may be found as explained in Fig 90. To obtain the points by which the ellipse may be determined, the circle of the base must be revolved until it is parallel to the front plane , it may then be divided as desired and the points revolved back to the line 1^F–5^F. As the circle revolves the points in it move in lines which in the front view appear parallel to the axis of the cone

In the top view the circle must be drawn when revolved until parallel to the top plane, the points marked in the circle of the front view can then be placed in that of the top view. As the circle revolves these points move, in

this view, in lines parallel to the axis of the cone, and these lines intersected by verticals from the points of the front view give the points in the ellipse.

The top view of *B* is the top view of *A* with its axis at an angle of 30°. Having drawn this top view the points of the front view may be found as explained in Art 112

FIG 92   *Front and top views of a cylinder whose axis is at 45° to the top plane and parallel to the front plane.*

Draw the front view first   Obtain the top view of the circles by means of points in the circles, as in Fig. 91

To place more than four points. draw a circle upon *1–2* of the front view and upon *3–4* of the top view; divide these circles as desired and so number the divisions that any point in the base is represented in both circles by the same figure or letter.

Project the points in the circle drawn in the front view to line *1F–2F*, and from this line project the points to the top view, where they will be situated in horizontal lines through the points of the circle there drawn. This amounts to revolving the circular base so that it is parallel first to the front plane and then to the top plane, as explained in Fig 91

FIG. 93.   *Front, top, and right side views of a square pyramid, whose axis is parallel to the front plane and at 45° to the top plane.   The edges of the base are at 45° to the front plane.*

The edges of the base, being at angles to both the front and the top planes, do not appear their real length in either view   One diagonal of the base is parallel to the front plane, the other diagonal is parallel to the top plane , each diagonal appears its real length upon the plane to which it is parallel, and thus the base of the pyramid, which is perpendicular to the front plane, is represented in the front view by a line whose length is equal to the diagonal of the base.   If the length of the edge of the base is given, that of the diagonal must be found by drawing the square, or one half of it, as at *A*   The front view should be drawn first, then the top or side view.

FIG 94.   (*A*) *Front, top, and right side views of a square plinth supporting a triangular prism.*

(*B*) *Top and front views of the same objects when the vertical faces of the plinth are at 30° and 60° with the front plane*

(*A*) First draw the views of the plinth , then draw the prism in the front view.   The distance *1–2* of this view is equal to the altitude of the triangular end of the prism.   To obtain *1–2*, supposing that the length of the side of the triangle is given, draw the real shape of the triangle as at *C*. From the front view of the prism obtain the other views

(*B*) Place the top view of *A* at the required angle, and obtain the front view, as explained in Arts 112 and 113

PLATE X.

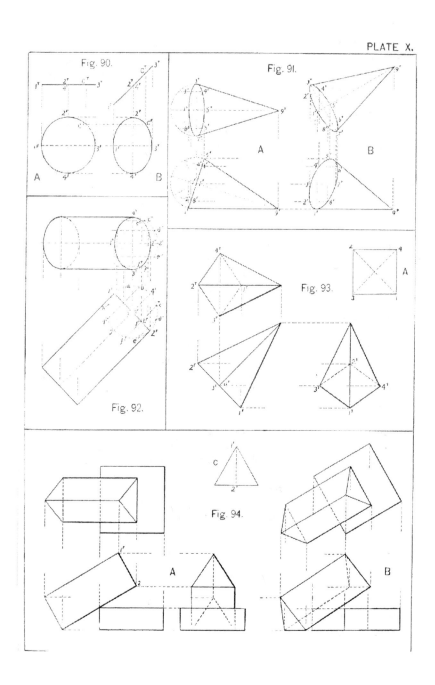

Fig. 90.

A

B

Fig. 91.

A

B

Fig. 92.

Fig. 93.

A

Fig. 94.

C

A

B

PLATE XI

FIG. 95   *Top, front, and right side views of an hexagonal pyramid intersected by a plane oblique to the base, the true shape of the section made by the plane, and the development of the lateral surface of the pyramid.*

All the points of intersection of the lateral edges with the cutting plane are seen in the front view, and may be projected to the other views. The width of the side view is the distance 3^T–5^T. The lateral edges 1–7 and 4–7 appear their real length in the front view, therefore the distance from the vertex, 7, to the points c and d of the section, may be measured from 7^F to c^F and from 7^F to d^F

To obtain the distances from the vertex 7 to the points of the section in the other lateral edges, draw, in the front view, horizontal lines through points a and b to the left or the right lateral edge, and measure the distances from point 7 upon these lines, as explained under Fig. 89

FIG. 96.   *Top and front views of a cone, showing the section made by a plane passing through its vertex.*

The line 1^F–3^F is the real length of the line represented by 1^T–a^T. Line 1^F–3^F is equal to the altitude, and line 2^T–3^T equal to the base of the isosceles triangle which is the section made by the plane   Art 161.

FIG 97   *Top and front views of a vertical cone intersected by a plane parallel to the front plane and in front of the axis of the cone, also the development of the lateral surface*

The circle of the base is cut in points 1 and 2, which are determined in the top view   To obtain 3, the highest point in the section, measure VT–3^T, the distance of the plane from the axis of the cone, and draw a vertical line $3'$–$3''$ the distance VT–3^T from the axis   This represents a side view of the plane, and intersecting the element of the cone at $3''$, gives the level of the highest point

To obtain two other points, intersect the cone by a horizontal plane B This gives a circle which appears a circle in the top view where it intersects the cutting plane in 4^T and 5^T, points in the required line of intersection , 4^F and 5^F must be in B^F and in projecting lines from these points of the top view.

Points 4 and 5 are in circle B, which is the base of a cone, just as 1 and 2 are in the base of the original cone. Any number of planes parallel to B may be taken, and each will give two points in the section, which is an hyperbola.

To develop the lateral surface of the cone proceed as explained in Art. 58 and under Fig. 88. To show the line of intersection upon it, measure, upon the arc, the distance of I^T and 2^T from b^T, and set this distance off from b^D in the development. The circle given by plane B becomes, in the development, an arc parallel with the one which bounds the lateral surface of the cone. In the circle of plane B are points 4 and $5;$ their distances from each other and from C are seen in the top view. These points will be placed in the development by measuring the distance $C\text{-}4^T$ and $C\text{-}5^T$ on the arc of the top view, and setting it off from C in the development. In this way all the points by which the section is obtained may be placed in the development.

PLATE XI.

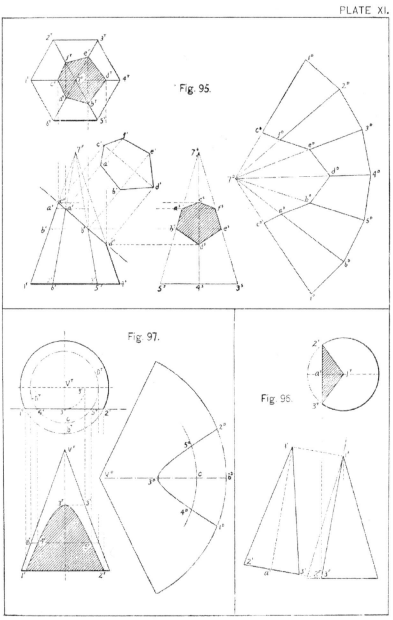

Fig. 95.

Fig. 97.

Fig. 96.

PLATE XII.

FIG. 98. *Top, front, and left side views of a vertical cone intersected by a plane at an angle with its base and cutting all the elements, the real shape of the section, also the development of the lateral surface of the cone*

The points in which the contour elements, V–1 and V–7, of the front view intersect the cutting plane are seen in the front view and may be projected to the top and side views The points f and g, in which the contour elements, V–4 and V–10, of the side view intersect the cutting plane are seen in the front view, and from this view may be projected to the side view. To obtain f and g in the top view the distance between these points may be measured in the side view and then set off in the top view , or a horizontal cutting plane may be taken through f and g, this gives a circle which in the top view gives f^T and g^T. To obtain other points in the section, other horizontal cutting planes may be used, as explained under Fig 97, or equidistant elements may be placed on the cone, and their intersections seen in the front view. As the surface is to be developed, the latter method is preferable. ' The lengths of these elements are found as if they were the edges of the pyramid of Fig. 95 The lateral surface should be divided into at least twelve equal parts.

To obtain in the development the direction of the curve at its termination on lines V^D–1^D, add one twelfth at either end of the arc which is the development of the base of the cone, and obtain lines V^D–$12'$ and V^D–$2'$ From V^D, on these lines set off the distance V^Db^D and V^Dc^D Trace the curve of intersection through these points.

FIG. 99. *Top and front views of a cone cut by a plane parallel to an element; also the real shape of the parabola which is the section.*

This section may be obtained as explained under Figs. 97 or 98 The drawing makes use of the cutting plane explained under Fig. 97.

FIG. 100. *Top, front, and right side views of a sphere, intersected by a vertical cone, and a horizontal cylinder*

In this figure the axes of the cylinder and cone pass through the centre of the sphere ; therefore the lines of intersection are circles. The planes of these circles are perpendicular to the front plane and are therefore represented by straight lines in the front view The circles in which the cone and sphere intersect are parallel to the top plane and are represented by circles in the top view

PLATE XII.

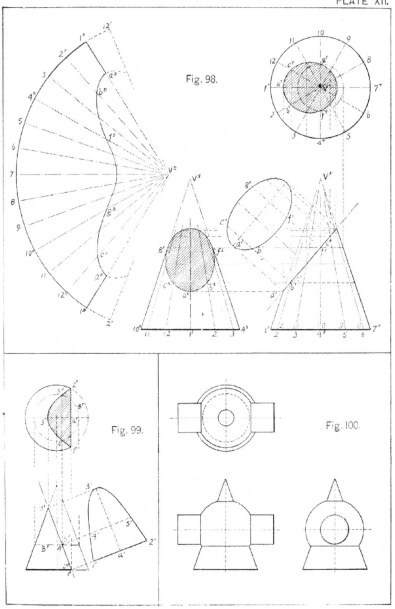

Fig. 98.

Fig. 99.

Fig. 100.

PLATE XIII.

FIG. 101   *Top and front views of an elbow extending from a conical support.*

If a cylinder or cone is cut by a plane oblique to its axis, one part may be revolved upon the other until the ellipses of the parts coincide in a second position   This will happen when an arc of 180° has been formed by the revolution of the parts   After such revolution the angle between the axes of the two parts is twice that of the angle of the cutting plane and the axis of the original solid.

The elbow illustrated may be formed by cutting a cylinder at 45° to its axis, and revolving the parts as explained

The plane of the section is perpendicular to the front plane and therefore the ellipse in which the vertical and horizontal cylinders intersect appears a straight line in the front view ; it appears a circle in the top view, for the surface of the vertical cylinder is perpendicular to the top plane

To develop the cylinder divide its surface into any number of equal parts by elements.   To do this in the case of the horizontal cylinder the circles must be revolved to appear their real shapes and divided as shown by the dotted lines   The elements having been drawn, their real lengths are seen in both views and may be set off upon the respective lines of the development.

When a cylinder is to be developed neither of whose bases is at right angles to the axis of the cylinder, it is necessary to assume a circle at right angles to the axis, this will develop into a straight line.   Thus, suppose the horizontal part to be represented in the front view by $(a\,b\,d\,e\,a)^F$.   To develop this cylinder, it will be necessary to develop a circle which is at right angles to the axis   This circle may be assumed anywhere, as at $a\,c$, and develops into a straight line a^Dc^D.   The development of the part at the right of the circle should be placed at the right of a^Dc^D, just as the part at the left is placed at the left of a^Dc^D.

The frustum of the cone is developed as explained under Fig 88

FIG 102   *Top and front views of an upright square prism inter-sected by a horizontal square prism at the right, and by a triangular prism at the left, also the developments of the lateral surfaces of the vertical and triangular prisms, and half that of the horizontal square prism*

The points in which all the edges of both horizontal solids intersect the surface of the vertical one are seen in the top view and are readily pro-jected to the front view

Develop the lateral surface of the upright prism as explained in Arts. 51 and 54. The horizontal faces of the square block intersect the upright prism in horizontal lines which are parallel to its bases, and which in the development must be parallel to *MN* and *OP*. The vertical faces intersect the prism in vertical lines parallel to the edges *B* and *D,* and the line of intersection of the horizontal and vertical square prisms on the development of the vertical prism is a rectangle whose length is the distance $1^T\text{-}c^T\text{-}3^T$; its width is the distance $1^F\text{-}2^F$.

The horizontal face of the triangular prism intersects the vertical square prism in lines which develop into a straight line parallel to *OP*, its length is the distance *7–5–6* of the top view; the positions of this line and also that of point *5* are seen in the front view.

The other developments require no special explanation In all the problems it is simply necessary to remember that the development gives the real length of every line of every surface and that the length of any line must be taken from the view in which it appears its real length, or, if not seen of its real length in any view, this must be determined as explained in Art. 114.

The development of any surface is most easily obtained by placing it so that one set of dimensions may be projected from the front view

PLATE XIII.

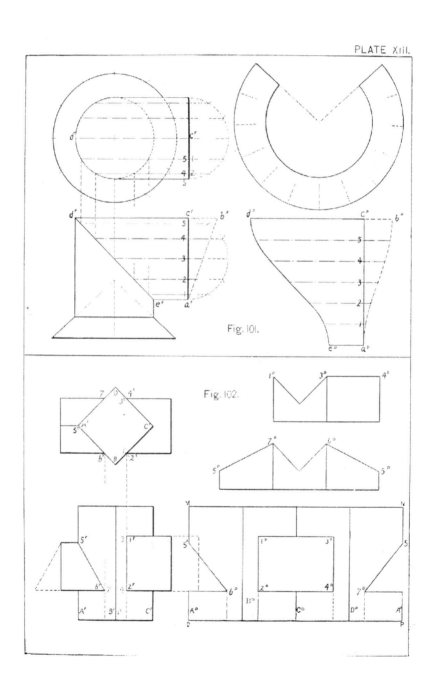

Fig. 101.

Fig. 102.

PLATE XIV.

FIG 103   *An upright square prism intersected at the left by a horizontal square prism, whose lateral faces are at 45° to the top and front planes, and at the right by an hexagonal prism with two of its lateral faces vertical, also the development of the lateral surface of the hexagonal prism, and half that of the horizontal square prism.*

The intersections are found as previously explained.

To develop the lateral surfaces of the penetrating prisms, measure the widths of the faces in the end views, and the lengths of the lateral edges in the front or top view, and combine these dimensions as explained

FIG. 104   *An upright square prism intersected at the right by a horizontal hexagonal prism, two lateral faces being horizontal, and at the left by a horizontal cylinder; also the developments of the lateral surface of the square prism, and half those of the hexagonal prism and of the cylinder*

In the front view, the intersections of the upper and lower elements *1* and *7* of the cylinder with the left edge of the prism are seen   In the top view, the intersections of the front and back elements *4* and *10* are seen. To obtain other points assume elements, to place which a side view is necessary, when determined in this view they may be transferred to the top view by revolving the circle in the top view until it appears a circle, dividing it, and numbering the points, to correspond with the end view; or projection methods may be used and the distances of the points *2*, *12*, etc., from a vertical line through the centre of the cylinder may be taken in the compasses from the side view and set off in the top view, one half each side of the axis.

The intersections of the elements and the prism are seen in the top view and may be projected to the front view

The intersections of the cylinder and the faces of the prism are semi-ellipses which appear a semi-circle in the front view

The intersection of the hexagonal and square prisms is found as explained

To develop the lateral surface of the square prism and show the lines of intersection upon it, first develop the entire lateral surface, then draw, in the front view, verticals upon the surface of the square prism through the points of the intersections   Place these lines in the development by setting off from A^D and C^D the distances, seen in the top view, of the lines from A^T and C^T; thus to obtain a^D and e^D draw a parallel to A^D at the distance $A^T a^T$ from it, and then set off in this line the distances of a^F and e^F from $A^F C^F$ In this way all the points necessary to determine the lines of intersection in the development may be obtained and the lines drawn through the points

PLATE XIV.

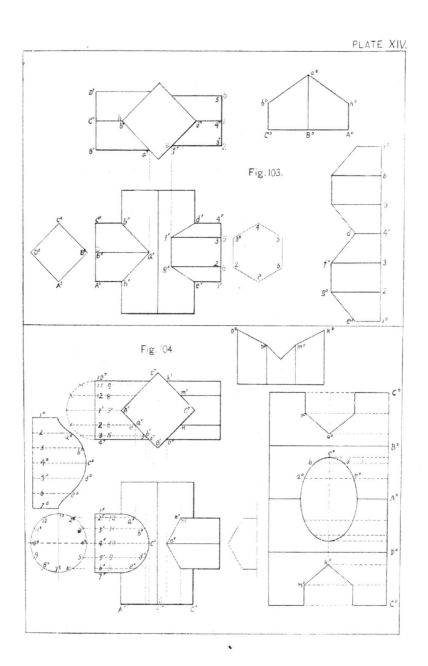

Fig. 103.

Fig. 104.

PLATE XV.

FIG. 105. *Top and front views of an upright cylinder intersected at the left by a horizontal square prism, and at the right by a horizontal cylinder; also the development of the lateral surface of the cylinder and half that of each intersecting body*

Both intersections and developments are the same in principle as those already explained. The line in which the square prism intersects the cylinder develops into a rectangle in the development of the cylinder, because the horizontal faces intersect the cylinder in circles which develop into straight lines, and the vertical faces intersect elements of the cylinder.

FIG 106. *Top and front views of a vertical cylinder intersected at the left by a square prism, and at the right by a triangular prism at 45° with the cylinder, also developments*

In the front view the distance $E^F F^F$ is equal to the distance $E'k$ of the view, $F'E'G'$, of the end of the prism, which must be drawn before the front view can be completed. It is not necessary to make this separate drawing, as the dotted lines in the front view which show half the end give all that is necessary

The points in which the lateral edges of the prisms intersect the cylinder are seen in the top view. The inclined sides of the prisms intersect the cylinder in ellipses, points in which may be found by assuming lines on the surfaces of the penetrating solids A line from I^T intersects the cylinder at a^T, from which the positions of a^F and c^F may be obtained A line from d^T gives the position of g^T. In this way any number of points may be obtained

The development of all the lateral surface of the triangular prism is given , point g is in a parallel to the lateral edges, and distant from E the distance EFd', point n is in a parallel to the lateral edges, whose distance from F is the distance $F^T e^T$

In the development of the square prism, points a^D and c^D are in parallels to the lateral edges, and the distances $B^{S-1}S$ and $B^{S-2}S$ from B^D.

In the development of the line of intersection upon the cylinder, point g^D is in a parallel to $f^D m^D$, whose distance from it is seen in the top view, from f^T to g^T, measured on the arc. Point n^D is in this same line, and h^D is in a parallel to $f^D m^D$, whose distance from it is the distance $f^T g^T h^T$, measured on the arc

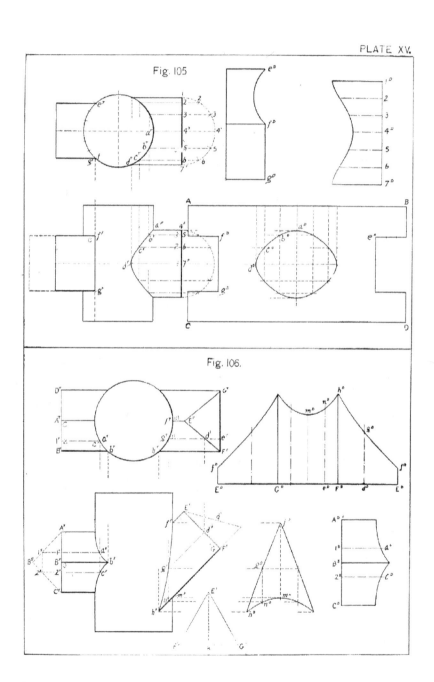

PLATE XV.

Fig. 105

Fig. 106.

PLATE XVI.

FIG. 107   *Front, top, and right side views of a horizontal square prism intersected by a vertical square pyramid; also developments.*

The intersections of lines *A V* and *C V* with the edges of the prism are seen in the front view, and the intersections of lines *B V* and *D V* with the prism are seen in the side view   Project these points of intersection from the side to the front view and from the front to the top view.   The intersecting surfaces are plane and therefore intersect in straight lines, which join the points of intersection

To obtain the lines of intersection in the development of the pyramid, measure the distances from the vertex *V* to the points of intersection in *VA* and *VC* in the front view, and the distances from the vertex to the points of intersection in *VB* and *VD* in the side view.   Set these distances off on the proper lines of the development and join the points

To obtain the points for the lines of intersection upon the lateral surface of the prism place points *k* and *d* in edge *2*, and *e* and *f* in edge *4*, their positions, in the edges, are seen in the front view   Points *a*, *h*, *g*, and *b* are in a plane perpendicular to the axis of the prism and must, in the development, be in a straight line perpendicular to the lateral edges   The distances of *a* and *h* from edge *2*, and of *b* and *g* from edge *4* are seen in the side view

FIG. 108.   *Top, front, and left side views of a horizontal cylinder intersected by an upright square pyramid; also developments.*

This problem is solved in the same way as Fig. 107, but the intersections are not straight lines.   Each lateral face of the pyramid cuts the cylinder in lines which are parts of the complete ellipse which will be given by the plane of each face cutting through the cylinder

The intersections of the lateral edges are seen in the front and side views   To obtain other points in the lines of intersection assume any line, as *V–1*, on the surface of the pyramid.   This line intersects the surface of the cylinder in points *b* and *g*; these points are readily obtained in the front and top views from *b^S* and *g^S*   In the side view *V–1* also represents the line *V–2* on the right face of the pyramid in which points *h* and *o* are situated. In this way any number of points in the curves may be found

To develop the pyramid and show the line of intersection, the distance from *V* to the points of the intersection in *VD* and *VB* must be measured in the front view ; the distances from *V* to the points of the intersection in *VA* and *VC* must be measured in the side view.

To place the points b, g and o, h in lines V-1 and V-2, draw through these points lines parallel to the base of the pyramid; the line through o intersects V^FBF in point 4^F. The distance of 4 from V is seen in the front view and 4^D is readily placed. A line through 4^D parallel to A^DB^D and intersecting V^D-2^D must give o^D, in this way all the points may be found

To obtain the lines of intersection in the development of the cylinder, measure, in the end view and around the arc, the distances apart of the elements containing the points; determine the positions of the points in the elements in the front view.

PLATE XVI.

Fig. 107.

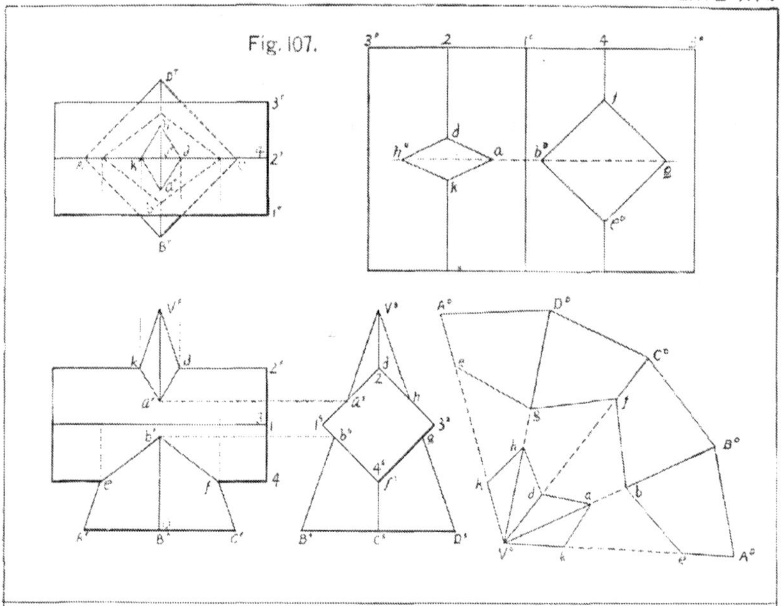

Fig. 108.

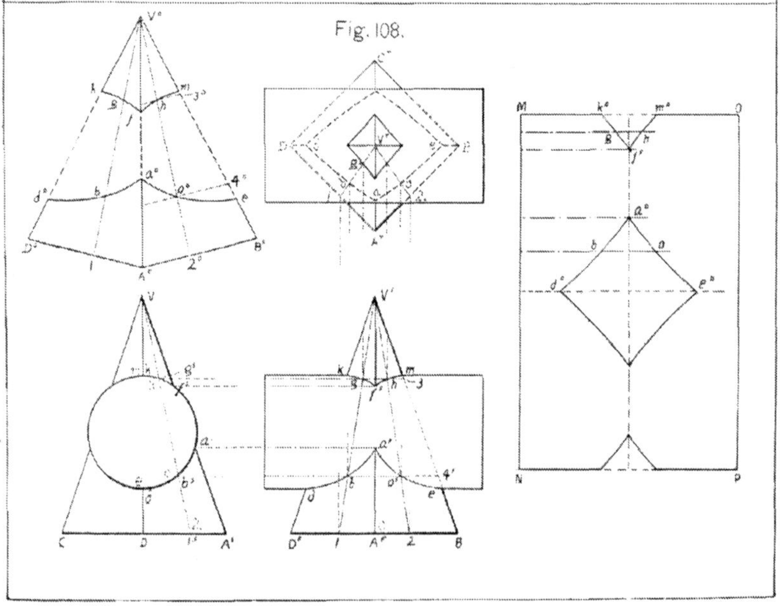

PLATE XVII.

FIG 109   *Top and front views of a vertical cone intersected at the left by a horizontal triangular prism, and at the right by a square prism*

The points in which edge D of the triangular prism and edges *2* and *4* of the square prism intersect the cone are seen in the front view, for these edges are in the plane of the contour elements   To find other points in the curves in which the cone is intersected by the lateral faces of the prisms, assume any horizontal cutting plane, as AB, this gives, upon the triangular prism, a rectangle whose width is *7–8*, and upon the square prism, one whose width is *5–6*, the plane intersects the cone in a circle C The circle and these rectangles intersect in points *c*, *d* and *a*, *b*, which must be points in the lines of intersection of the cone with the triangular and square prisms. As many points as are desired may be obtained by means of other cutting planes.

FIG 110.   *Top and front views of a vertical cone intersected by a horizontal square pyramid.*

Edges E and F intersect the contour elements of the cone. To find the intersections of edges G and H. and also to obtain other points in the intersections, intersect the solids by cutting planes. CD is a horizontal cutting plane and gives four points, *e, f, g, h,* in the lines of intersection

FIG. 111   *A vertical cylinder intersected by a horizontal cone, also developments*

The principles involved have been previously explained

To develop the cone, find the distances of points *b, c, d, e,* and *f* from the vertex V, by measuring upon the contour elements, as explained under Fig. 98.

Obtain the line of intersection in the development of the cylinder by measuring, in the top view and around the arc, the distances between the elements containing the points, and by measuring, in the front view, the positions of the points in these elements.

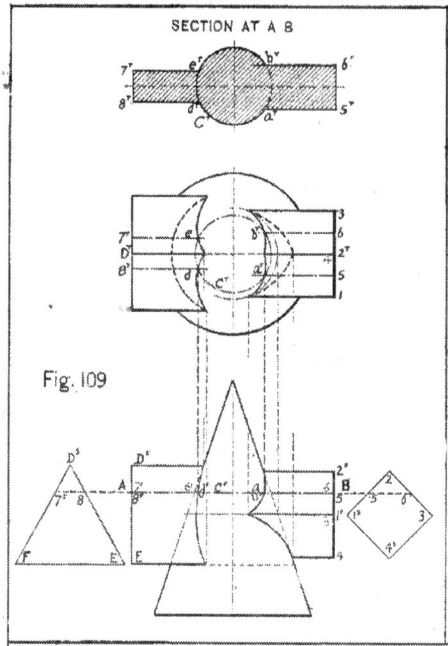

PLATE XVII.

SECTION AT A B

SECTION AT C D

Fig. 109

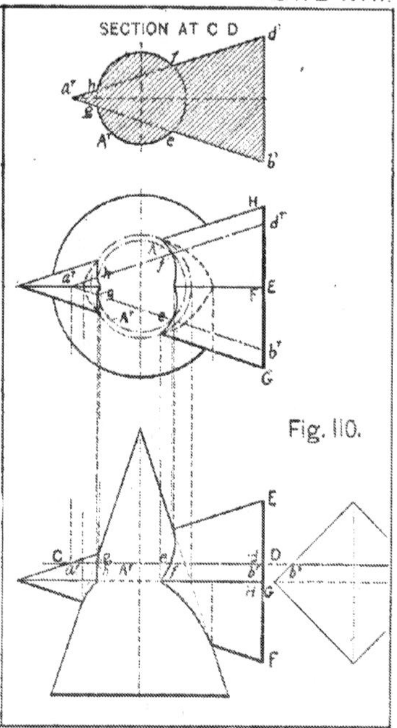

Fig. 110.

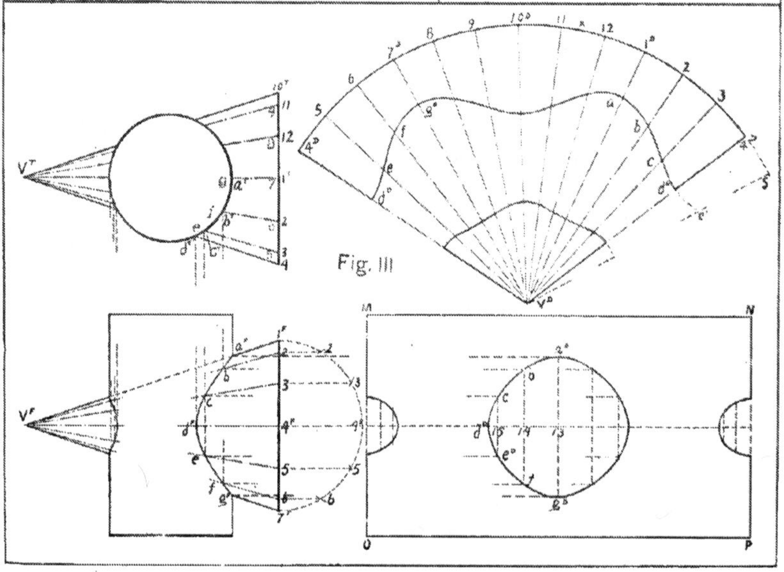

Fig. III

PLATE XVIII.

FIG. 112. *Front, top, and left side views of a loose joint hinge; also front and top views of the two parts forming the hinge, when they are separated from each other*

FIG. 113. *Front, top, and left side views of a sash lift; also a section on AB.*

FIG. 114. *Front, bottom, and left side views of a drawer pull; also a section on CD.*

Most of the objects shown on Plates XVIII to XXII inclusive do not require all the views given to make them complete working drawings. In Fig. 113 or 114, for instance, simply the front and top views and the section furnish all the information necessary to make the object The side views are given to familiarize the student with the arrangement of all the views.

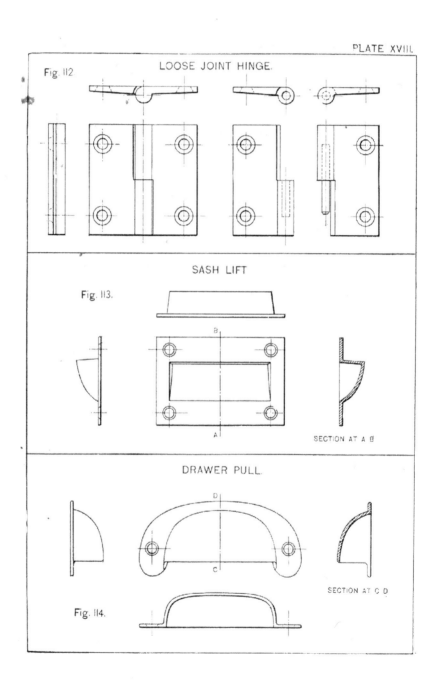

PLATE XVIII.

LOOSE JOINT HINGE.

Fig. 112.

SASH LIFT

Fig. 113.

SECTION AT A B

DRAWER PULL.

SECTION AT C D

Fig. 114.

PLATE XIX.

FIG. 115   *Front, top, and right side views of an iron-cased bolt, with details of the same.*

Dotted lines are very confusing when numerous ; therefore when many would be required it is customary to represent the different parts of an object separately.

In this figure all the views are needed to make the construction clear.

The ornamental tracery upon the object is not represented.   Such details should be omitted when pupils make working drawings of common objects.

PLATE XIX.

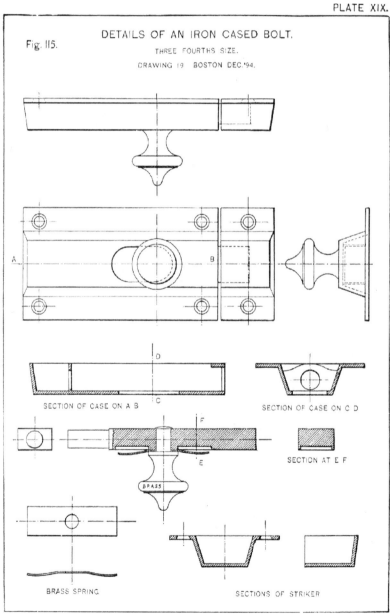

DETAILS OF AN IRON CASED BOLT.

Fig. 115.

THREE FOURTHS SIZE.

DRAWING 19   BOSTON DEC. '94.

SECTION OF CASE ON A B

SECTION OF CASE ON C D

SECTION AT E F

BRASS

BRASS SPRING

SECTIONS OF STRIKER

PLATE XX.

FIG. 116.   *Details of an iron side pulley*
A is a front view of the pulley.
B is a side view of the pulley.
C is a section of the frame on GH.
D is a back view of the frame.
E is a front view and a section on EF of the wheel.
FIG. 117.   *Front, top, and side views of an iron bracket.*
FIG. 118.   *Front, top, and bottom views of an iron clamp.*

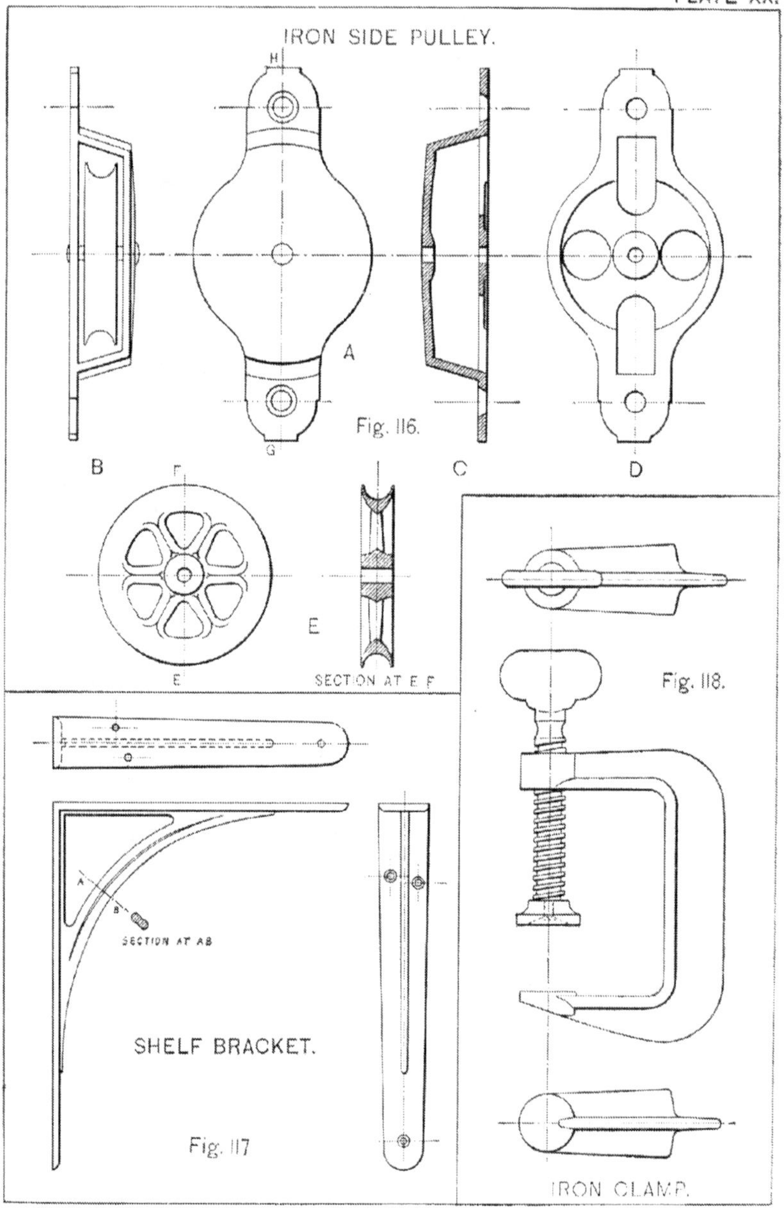

PLATE XX.

IRON SIDE PULLEY.

Fig. 116.

B C D

E

SECTION AT E F

Fig. 118.

SHELF BRACKET.

SECTION AT AB

Fig. 117

IRON CLAMP.

PLATE XXI.

Fig. 119. *Free-hand sketches of an iron caster, from which finished drawings are to be made.*

In making such sketches the proportions of all the parts should be represented as truly as possible ; but the parts should be drawn free-hand, and the drawings completed before the object is measured. It is not necessary that the proportions be exact, but it is better to have them so, as errors in measuring or in figuring the sketches may be made. If the sketches are correct in their proportions, these errors will be shown when the drawings are made to scale, and compared with the free-hand sketches.

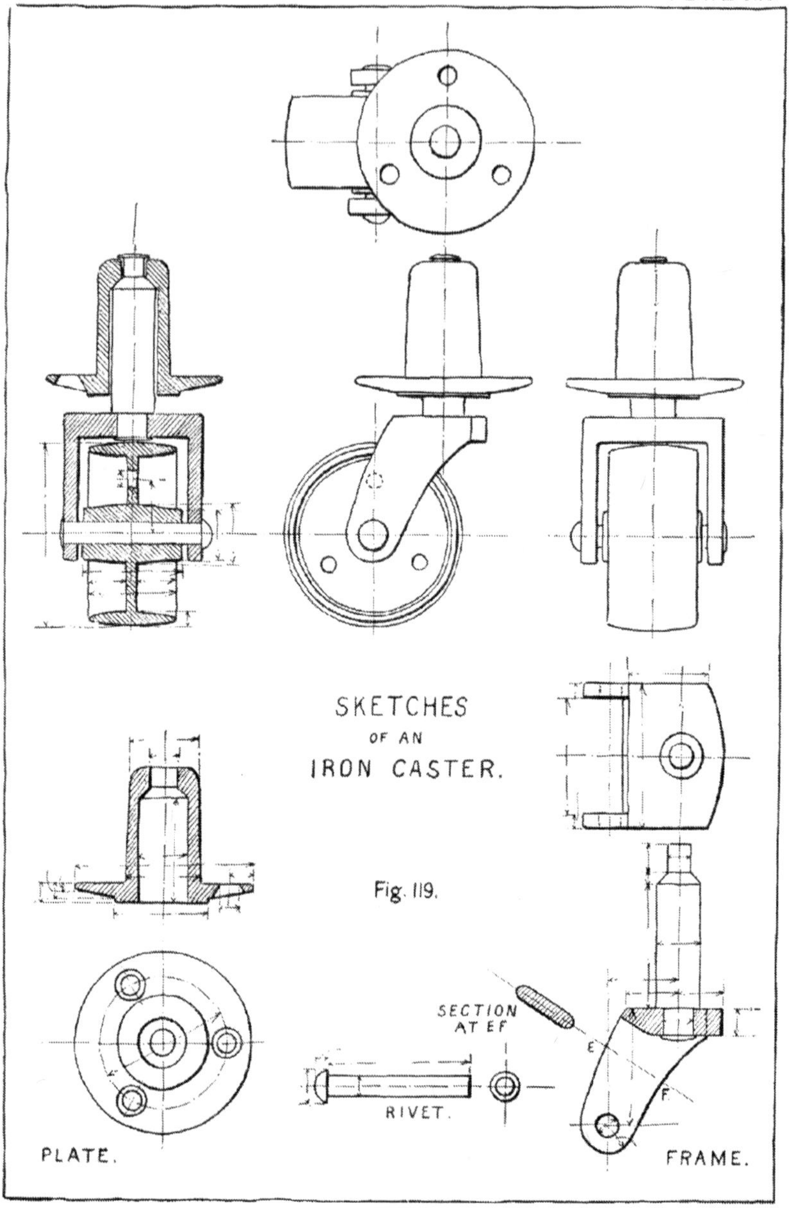

PLATE XXI.

SKETCHES
OF AN
IRON CASTER.

Fig. 119.

PLATE.

RIVET.

SECTION
AT EF

FRAME.

PLATE XXII.

FIG. 120.   *Views of a wooden faucet*

A is the front view.

B is the top view.

C is the left side view.

D is a section at KL.

E is a section at MN

The plug is surrounded by cork, which gives a tight joint   The cork is shown in the sections by the dotted parts

The cut surface of the wood may be shown as illustrated on page 61.

The lines in which the conical parts of the faucet intersect the faces of the octagonal part are the only points requiring explanation ; these lines are really hyperbolas (Fig. 97).   It is customary to represent such curves by arcs of circles of which the highest points and the points in the edges are found exactly ; the lines of intersection upon the narrow faces are so slightly curved that straight lines may be, and generally are, substituted for the arcs

To determine the points for the arcs, notice that, in the front view, the contour elements of the conical parts must intersect the upper and lower lines at the central or highest points of the curves in two wide faces   As the wide faces are of equal width, the highest points in the curves of these faces, in both front and top views, will be in vertical lines drawn through the points of intersection.   To find the points in the edges, an edge must be represented so that its real distance from the centre of the faucet shall be seen ; to do this revolve, about the centre of the end view, point a^S into the position a'; project from a' a horizontal line which represents an edge. and intersecting element 2-1 of the cone at a'', gives the position of the point where the conical surface cuts the edges of the faucet   These edges are equally distant from the centre of the faucet ; therefore in both front and top views all the arcs must end at the edges in points which are in a vertical line drawn through a''.

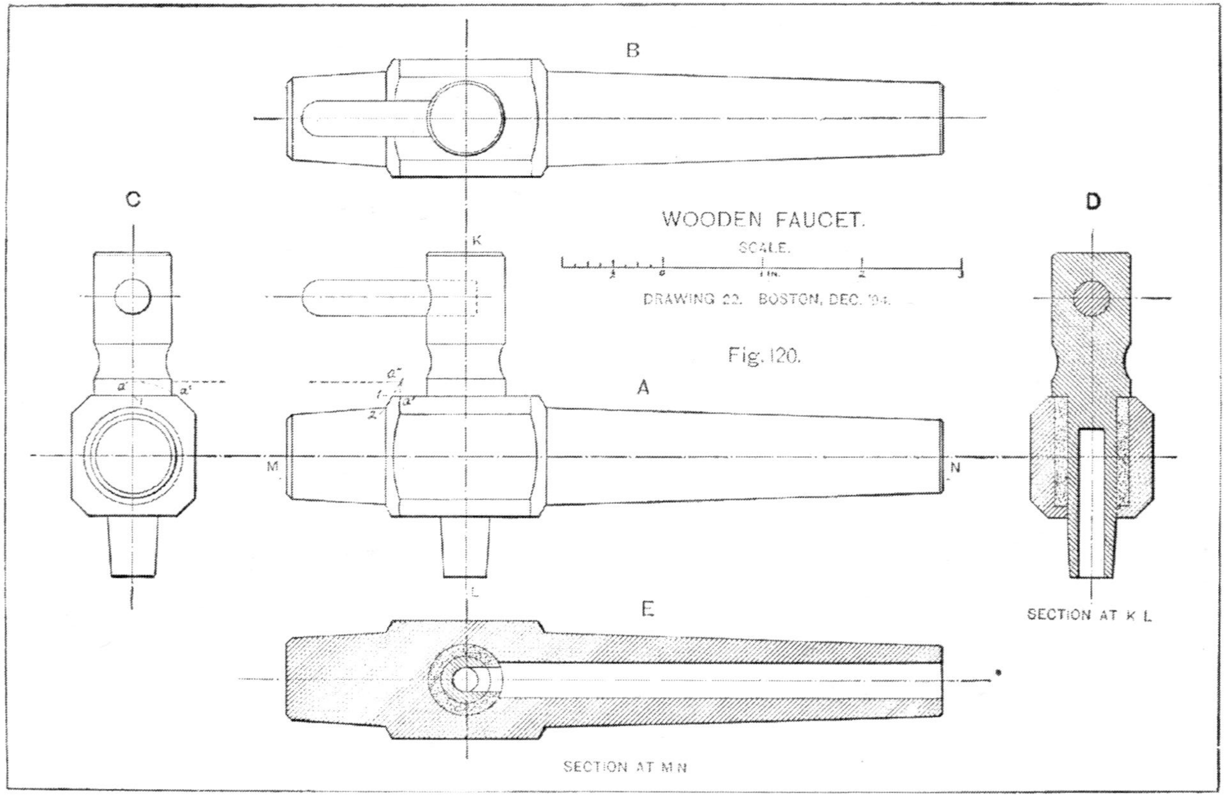

PLATE XXII.

D

SECTION AT K L.

B

WOODEN FAUCET.

SCALE.

DRAWING 22. BOSTON, DEC. '96.

Fig. 120.

A

K

L

E

SECTION AT M N

C

PLATE XXIII.

FIG 121. *A front view and section of a washer*

FIG. 122. *Front and top views of the end of a rod adapted to hold a lever, which is pivoted on screws inserted one in each part of the fork*

The forked part is cast and is screwed to the end of a round rod, of which a part is shown.

The information which an end view would give is necessary to make the drawings give all the facts of form but the placing of the letter D after the dimension which shows the diameter of the round part of the fork, makes the end view unnecessary

Threaded holes are best shown as illustrated ; in one view two circles are drawn, the outer being dotted and the inner full. Some draughtsmen represent part of the outer circle by a full line which becomes tangent to the inner circle   This is more as the thread appears, but the simpler representation is the better.   In the other view the dotted lines which show the section of the thread are at an angle of 60° with each other , they should not join each other, as this will destroy the effect of a dotted line , they should be drawn free-hand with a writing-pen.   Only the outline of the thread should be represented, for dotted lines in place of all the full lines shown on Plate XXVI will give an unsatisfactory drawing

FIG. 123. *A top view and a front view one half of which shows a section through the centre of a machine detail called a gland, and used for compressing packing about a piston to prevent the escape of steam or water.*

It is customary to show in section one half (or any other part) of the view of a symmetrical object : this saves the time required to make complete views of the outside and of the inside of the object

In any working drawing no more lines should be given than are necessary to show the construction ; dotted lines are especially to be avoided

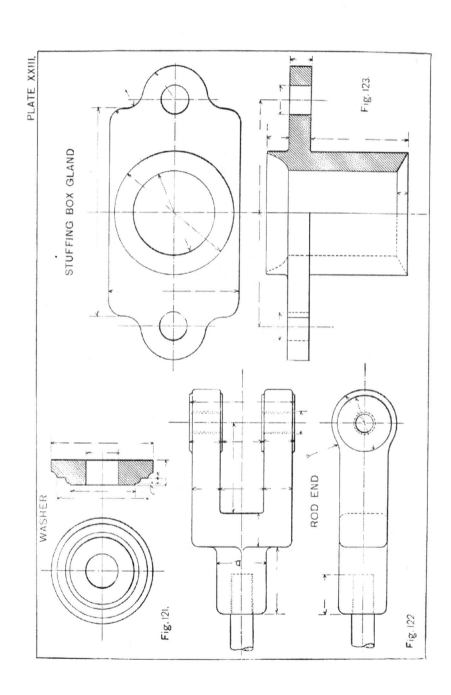

PLATE XXIII.

STUFFING BOX GLAND

WASHER

Fig. 121.

ROD END

Fig. 122

Fig. 123

PLATE XXIV.

FIG 124. *Top and side views, and a front view, one half of which shows a section, of a bearing for the end of a shaft which moves back and forth in a plane, and so requires a bearing that rotates   The shaft rests upon a box of composition B, which is replaced when worn.*

FIG. 125. *A front view, of which one half is in section, and a vertical cross section, of a bearing used upon a locomotive.*

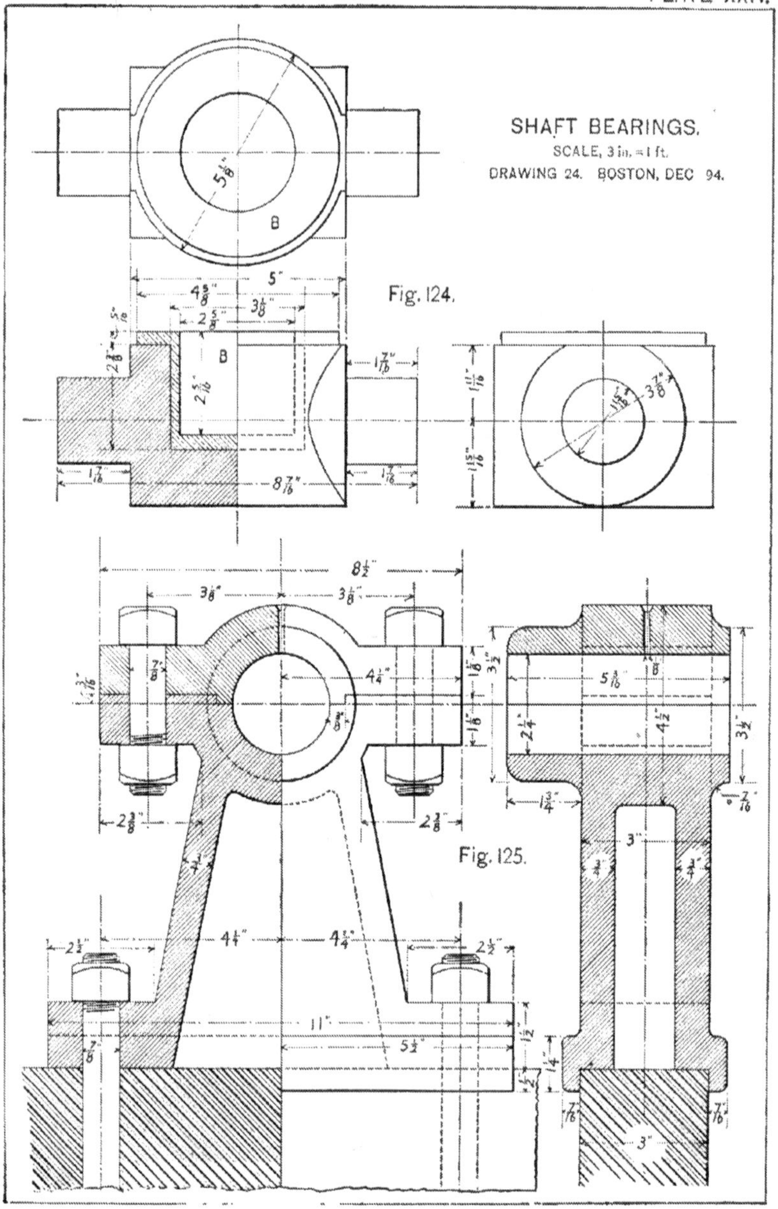

PLATE XXIV.

SHAFT BEARINGS.
SCALE, 3 in. = 1 ft.
DRAWING 24. BOSTON, DEC 94.

Fig. 124.

Fig. 125.

PLATE XXV

THE HELIX.

If a point moves in two directions about a given line as axis, — that is around and along the line at the same time, — a helical curve results.

The simplest form of this curve is found in a common spring   The lines which bound the threads of screws are helices ; these curves are found in many constructions

The motion of the point in its different directions may be uniform or variable ; in the common forms of the curve both motions are uniform, and are produced by a point which moves uniformly around and along a cylinder at the same time

The distance which the point travels along the cylinder, in going once around it, is called the *pitch.*

To draw the common helix, divide the circle which is a view of the line in which the generating point moves about the axis, into any number of equal parts, and divide the pitch into the same number of equal parts. When the point has moved upon the circle of the end view over one of the equal spaces into which the circle is divided, it has moved in the other view, along the axis from its starting point, a distance equal to one of the equal spaces into which the pitch is divided ; when it has moved one quarter around the circle it has moved one quarter of the pitch ; when it has moved one half around the circle it has moved one half of the pitch, and so on for the whole revolution of the generating point   Hence to obtain all the points for the view of the helix, draw parallels to the axis from the points of division in the circle, and intersect these lines by perpendiculars to the axis from the equal divisions of the pitch   If the points of the circle and the pitch are numbered in order from the first point, the intersections of lines of the same number will be points in the required curve

FIG. 126.   *A helical curve upon a cylinder.*

The curve is found as explained above.   The division of the pitch into twelve equal parts is advisable for pupils.   To avoid error, the points of the bottom view and also those of the pitch may be numbered from 1 to 12. The curve is symmetrical , its vertices *12* and *f* must be curved.

If the cylinder is developed, the helix will become the hypothenuse of a right-angled triangle *A*, of which one side of the right angle is equal to the circumference of the cylinder and the other is equal to the pitch.

When many curves of the same pitch are desired, a templet should be

made of thin wood to fit the first one drawn; by this the others can be quickly and accurately drawn

FIG. 127. *A cylinder with a helical blade or surface, formed by the revolution of a line perpendicular to the cylinder around and along the cylinder at the same time.*

The two ends of the generating line describe helices of the same pitch, but on cylinders of different diameters Each helix is found as already explained.

Instead of revolving from left to right the line revolves from right to left, and the helices advance along the cylinder in the opposite direction to that of the helix in Fig. 126

Threads are called *right-handed* when they advance as in Fig. 126, and *left-handed* when they advance as in Fig. 127. The most common examples of both kinds of threads are found on the axles of all wagons The threads on the right hand ends are right-handed; the threads on the left hand ends are left-handed, that is, the nuts are screwed upon them by the motion which unscrews those on the right hand ends.

A helical curve may be generated by a point which moves along the surface of a cylinder at a varying rate of speed, while its motion around the cylinder is uniform; or it may be generated by a point whose motion in both directions is variable. A helical curve may be traced upon a conical or spherical surface A helical surface may be generated by a line which moves as in Fig. 127, or by a line which is inclined to the axis of the cylinder. These problems are of too advanced a nature to be given in this book

FIG 128 *A V-threaded bolt and nut, the upper part of the bolt and the nut showing a section through the centre*

The section of this thread is an equilateral triangle. The size of the thread depends upon the diameter of the bolt; there are regular standard threads for bolts of all diameters, and other standard threads for pipes. Iron bolts of the following diameters have threads as specified

Diameter of Bolt		Threads per inch
¼ inch	. . .	20
⅜ "	. . . .	16
½ "	. . .	12
¾ "	. .	10
1 "	. . . .	8
1½ "	. . .	6

We will suppose the screw shown to be of wood; the thread is therefore large in proportion to the bolt. First draw the section of the lower thread and divide the pitch into twelve equal parts As only half of the

bottom view is shown, this is divided into six equal parts. The outer and inner helices are obtained as explained under Fig. 126, and the drawing of the bolt is completed by drawing the section of the thread which connects the outer and inner helical curves

In exact drawings of bolts whose threads are large, the *V*-shaped outline of the section of the thread will come inside the correct projection of the helical surfaces forming the thread , but to obtain the exact projection is so complicated a problem that generally the thread is represented as illustrated. A straight line tangent to the two helices approximates the actual projection. As the helices are not angular at their vertices it is sometimes necessary to represent the thread in this way instead of by the line of the section which is given in the figure

When the first outer and inner helices are complete, the others may be obtained by setting off the pitch, with the dividers, from the points of the first lines as many times as other curves are desired.

The lines of the thread in the interior of the nut have the opposite direction to those upon the bolt ; they correspond to the dotted lines of the bolt and are obtained in the same way.

PLATE XXV.

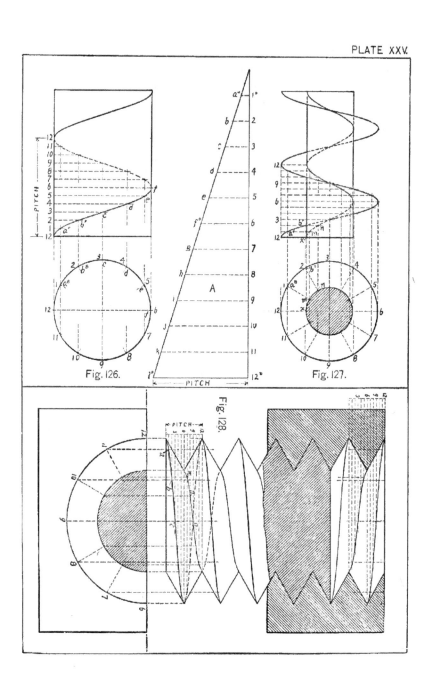

Fig. 126.

Fig. 127.

Fig. 128.

PITCH

PLATE XXVI.

FIG. 129   *A square-threaded bolt and nut, the upper part of the bolt and the nut being in section.*

In the square thread both the thread and the space are square, each occupying one half the pitch   There are two helical curves at the inside of the thread, and two at the outside ; the points for these curves are found as already explained

Figs 128 and 129 give the actual projections of the common forms of threads.   These threads are cut in a lathe, and drawings for them are un-necessary, as all that the workman requires is to know the diameter of the bolt, the pitch of the thread, and its shape.   These facts might be written upon the drawing, but it is the custom to represent threads conventionally by means of straight lines, which require less time to draw than do the curves.

FIG 130.   *A bolt with V and square threads, and a nut in section for each.*

At the right a *V*-threaded bolt and nut are shown   The drawing is given by substituting straight lines for the curves of Fig. 128.

Two methods of representing the thread of a square-threaded bolt and nut are shown at the left.   The threads at the left end of the bolt and in the nut are obtained by substituting straight lines for the curves of Fig 129. The threads near the centre of the bolt are shown more simply by straight lines representing only the outer edges of the thread

FIG 131   *Three different ways of representing a V thread.*

At *A* is shown a method suitable for use when the screw is very small , the lines of the bolt in such a case may be omitted, especially when the end of the bolt shows for a little distance outside of its nut.

At *B* is shown the representation used by many draughtsmen.   Some-times the long lines are made heavy instead of the short lines   Heavy short lines give the best effect of the thread.

At *C* is shown a satisfactory representation, which requires less time to produce than the other methods.

It is not necessary that there be as many long lines per inch as there are threads, though it is well for pupils to represent each thread by a long line, because they are not able to space by eye so as to produce a satisfac-tory result

The angle at which the lines of these conventional representations are drawn, should be determined by the spacing of the lines.   In all common

threads point *3* of Fig. 131 should be half-way between points *1* and *2*. This drawing represents correctly a single-threaded right-handed bolt. The opposite direction of the long line, that is from *3* to *2*, will represent a left-handed thread.

FIG. 132.   *A triple-threaded bolt.*

A bolt may have any number of parallel threads.   In this case there are two threads, *B* and *C*, between the turns of the thread *A*.   One turn of the bolt moves it the distance of the pitch *A A;* the threads *A. B*, and *C* give three times the strength that would be given by thread *A*.   To draw this bolt, divide the pitch into as many equal parts as there are threads, and place one thread in each space, as in a single-threaded bolt.

FIG. 133 shows the conventional shading sometimes added to a drawing representing cylindrical objects.   Part of Fig. 132 shows how a vertical cylinder is shaded.

PLATE XXVI.

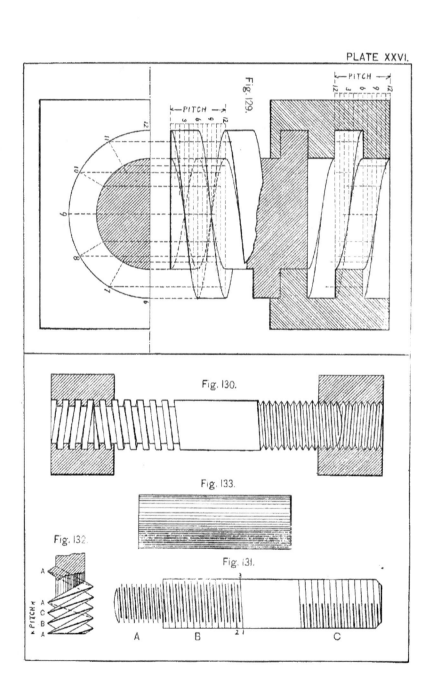

Fig. 129.

PITCH

PITCH

Fig. 130.

Fig. 133.

Fig. 132.

Fig. 131.

PITCH

A

A
C
B
A

A   B   C

PLATE XXVII.

Fig. 134.   *Three views of a square bolt-head.*

Fig. 135.   *Three views of a square bolt-head, whose corners have been chamfered (bevelled).*

The surface of the chamfer is really that of a cone, which is shown by the dotted lines.   This cone will be intersected by each of the four faces of the head in an hyperbola.   In practice this is never drawn, but is represented by an arc of a circle, whose points are found as follows .   The contour elements of the cone intersect the vertical lines which represent the right and left sides of the head, and give the highest points of the curves of intersection.   To obtain the lowest points, which must be in the edges since they are farther from the centre than any other lines of the faces, revolve 2^T to $2'$ and draw from $2'$ a vertical line to intersect the element of the cone at $2''$ This vertical represents an edge of the head, when its real distance from the centre of the bolt is seen, and $2''$ must be the point in which the edge is cut by the chamfer.   The highest points in the curves of intersection must be at the same level upon all the faces ; the lowest points must also be at one level , thus the arcs which represent the chamfer must be tangent to a horizontal line drawn through 1 and must end in the edges in points in a horizontal line drawn through $2''$.

The angle of the chamfer ranges usually from 30° to 45°, and much or little may be cut from the head , thus the circle left upon the top may be tangent to the sides, or smaller than in Fig. 135.

Fig. 136.   *Three views of a square bolt-head with a spherical top.*

The top is a portion of a sphere, and its section by any one of the faces of the bolt-head is an arc of a circle.   The highest point is seen at 1^F, and the lowest point is obtained as in Fig 135

The radius of the spherical top is usually about twice the diameter of the bolt.

Fig. 137.   *Three views of a square bolt-head when its vertical faces are at an angle of 45° with the front plane*

The arcs of circles at the top of each face appear ellipses but are represented by arcs of circles.   The lowest points in the arcs are in the edges and are seen in both views   To obtain the level of the highest point revolve the centre of any face, as 3^T, to $3'$ and project to the front view as explained in Fig. 135.   The highest points are in the centre of each face , their level is thus given at $3''$·   The arcs of both views must be contained between horizontal lines through 1^F and 3^F, as in Fig 135.

FIG. 138. *Three views of a square nut so placed as to show two faces in the front view.*

The nut is shown upon a bolt, which should always be represented as projecting slightly beyond the nut. The points for the curves representing the chamfer are found as has been explained.

A square nut or bolt-head should be so placed as to show one face only, as in Fig. 135.

The thickness of the square nut and bolt-head is equal to the diameter of the bolt. The distance between the parallel sides varies ; it is often 1½ times the diameter of the bolt plus ⅛ of an inch.

FIG. 139. *Three views of an hexagonal bolt-head.*

PLATE XXVII.

Fig. 134.

Fig. 135.

Fig. 136.

Fig. 137.

Fig. 138.

Fig. 139.

PLATE XXVIII.

FIG. 140.   *Three views of a bolt having an hexagonal head with a spherical top.*

The lowest points in the curves of the chamfer, which are really arcs in this figure, are in the edges and are determined in the front view; the highest points are in the centres of the faces and are determined in the side view The arcs which are drawn to represent the chamfer must be tangent to a line drawn through *2*, and have their lowest points in a line drawn through *1*.

FIG. 141.   *Three views of a bolt, with an hexagonal head and nut, whose corners are chamfered.*

The circle on the top of the head given by the chamfer appears, in the front and in the side view, a straight line whose length is equal to the diameter of the circle. Drawing from the extremities of these lines the elements of the cone (generally at 30°) determines, in the front view, the lowest points in the curve by the intersections of these lines with the edges; and the highest points are determined in the side view, where the lines are seen intersecting the centres of the faces. In this figure, and also Fig. 140, the arcs must end in the edges at points given by a line drawn through point *1*, and must be tangent at the centre of each face to a line drawn through point *2*. The curves of the chamfer in this figure are hyperbolas, as they are in Fig. 135.

The thickness of the hexagonal nut and bolt-head is equal to the diameter of the bolt, the distance between the parallel sides varies It is not necessary to represent the exact proportions; therefore many draughtsmen make the conventional drawings shown in Figs 139, 140, and 141. in which the long diagonal of the head, or nut, is made twice the diameter of the bolt. This is larger than the standard sizes, but is the best representation for practical drawings.

When only one view of an hexagonal bolt-head or nut is shown, it should represent three faces of the object

The draughtsman does not need to find the exact points for the curves of the chamfer, as he knows the appearance and can give it by eye; but it is well that students should know how to obtain the correct points.

FIG. 142.   *A spring or a pipe of helical form.*

Make the drawing of the helical curve, which is the centre line of the spring, as explained in Fig 126. With different points in the curve as centres, draw circles whose diameters are equal to that of the spring. The outline of the spring must be drawn tangent to these circles.

PLATE XXVIII.

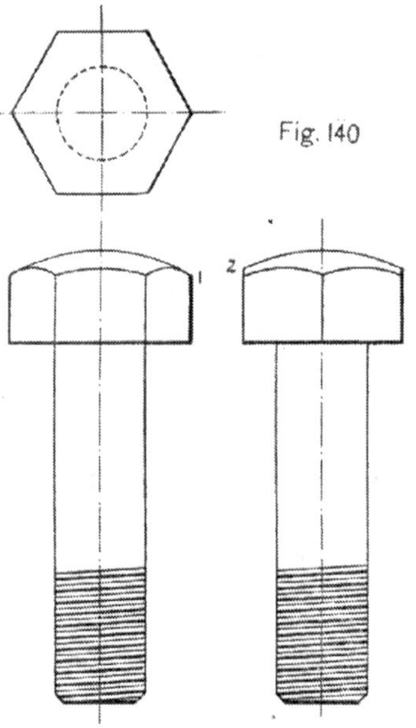

Fig. 140

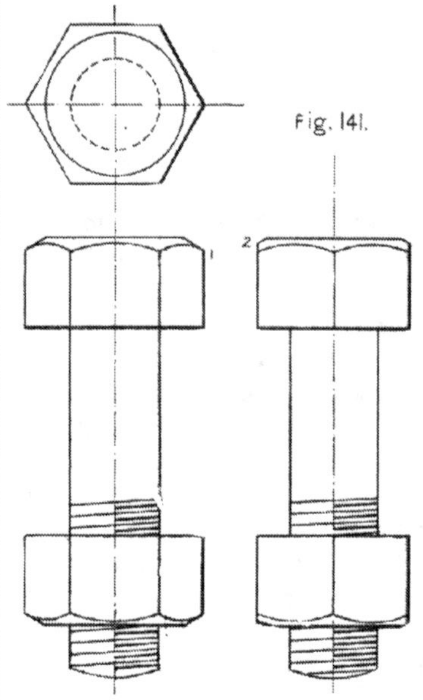

Fig. 141.

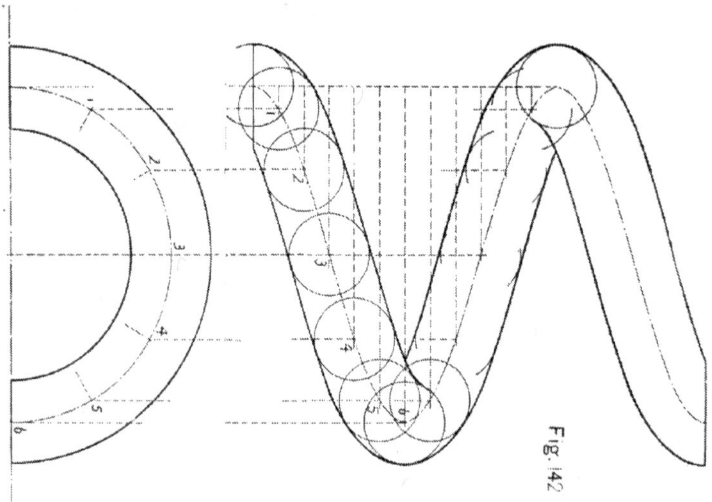

Fig. 142

DEFINITIONS.

Altitude. The perpendicular distance between the bases, or between the vertex and the base, of a solid or plane figure.

Angle The difference in direction of two lines which meet or tend to meet. The lines are called the *sides*, and the point of meeting, the *vertex* of the angle.

An angle is measured by means of an arc of a circle described from its vertex as a centre and included between its sides The centre of the arc is the vertex of the angle.

If the radius of the circle moves through $\frac{1}{360}$ of the circumference, it produces an angle which is taken as the unit for measuring angles, and is called a *degree*.

The degree is divided into sixty equal parts called *minutes*, and the minutes into sixty equal parts called *seconds*.

Degrees, minutes, and seconds are denoted by symbols. Thus 5 degrees, 13 minutes, 12 seconds, is written 5° 13′ 12″.

A RIGHT ANGLE is one which is formed by the radius moving through $\frac{1}{4}$ of the circumference. It is an angle of 90°. A *straight* angle is formed when the radius has moved over $\frac{1}{2}$ of the circumference. It is an angle of 180°.

ACUTE ANGLE. An angle less than a right angle.

OBTUSE ANGLE. An angle greater than a right angle.

OBLIQUE ANGLE. One which is not a right or a straight angle.

REFLEX ANGLE. One which is greater than 180°.

ADJACENT ANGLE Two angles are adjacent when they have the same vertex and a common side.

DIHEDRAL ANGLE. The opening between two intersecting planes

SOLID ANGLE One formed by planes which meet at a point

Apex. The summit or highest point of an object.

Arc. See Circle.

Axis of a Solid An imaginary straight line passing through its centre and about which the different parts are symmetrically arranged.

Axis of a Figure A straight line passing through the centre of a figure, and dividing it into two equal parts.

Axis of Symmetry. A straight line so placed in a solid or a plane figure that every straight line meeting it at right angles and extending in each direction to the boundary of the solid or figure is bisected at the point of meeting. In many solids and plane figures an axis of symmetry cannot be drawn

Base The opposite parallel polygons of prisms. The polygon opposite the vertex of a pyramid The plane surfaces of cylinders and cones. The opposite parallel sides of a parallelogram or trapezoid The shortest or longest side of an isosceles triangle, and any side in any other triangle, but usually the lowest.

Bisect To divide into two equal parts.

Bisector A line which bisects.

Cinquefoil. A figure composed of five leaf-like parts.

Circle A plane figure bounded by a curved line, called a circumference, all points of which are equally distant from a point within called the *centre*.

The boundary line is called the CIRCUMFERENCE

DIAMETER A straight line drawn through the centre, and connecting opposite points in the circumference, as *a b*

RADIUS. The distance from its centre to the circumference, as *c e*

SEMI-CIRCLE. Half a circle, formed by bisecting it with a diameter, as *a d b a*

ARC. Any part of the circumference, as *e b*.

CHORD. A straight line whose ends are in the circumference, as *f g*

SEGMENT. The part of a circle bounded by an arc and a chord, as *f h g f.*

SECTOR. The part of a circle bounded by two radii and an arc, as *b e c b*

QUADRANT A sector bounded by two radii and one fourth of the circumference, as *a c d a*

TANGENT. A straight line which meets a circumference, but being produced does not cut it, as *k d* The point of meeting is called the *point of contact* or *point of tangency.*

Circumscribe. A polygon is said to be circumscribed about a circle when each side of the polygon is a tangent to the circle , and a circle is said to be circumscribed about a polygon when the circumference of the circle passes through all the vertices of the polygon.

Concave. Curving inwardly.

Cone. A solid bounded by a plane surface called the *base,* which is a circle, ellipse, or other curved figure and by a lateral surface which is everywhere curved, and tapers to a point called the *vertex* Its base names the cone. Thus a circular cone is one whose base is a circle

A RIGHT CIRCULAR CONE is generated by an isosceles triangle which revolves about its altitude as an axis The equal sides of the triangle in any position are called *elements* of the surface. The length of an element is called the *slant height* of the cone. Unless otherwise stated "cone" means a right circular cone.

A FRUSTUM OF A CONE is the part included between the base and a plane parallel to the base and cutting all the elements of the cone

A TRUNCATED CONE is the part included between the base and a plane oblique to the base and cutting all the elements of the cone.

Concentric. Having a common centre.

Conic Section. A section obtained by cutting a cone by a plane.

Construction The making of any object

Construction Lines. The lines by which the desired result is obtained

Constructive Drawing. A drawing intended for the workman who is to make the object.

Contour. The outline of the general appearance of an object

Contour Element An element which is in the contour of an object.

Convergence Lines extending toward a common point, or planes extending toward a common line

Convex. Rising or swelling into a spherical or rounded form

Corner The point of meeting of the edges of a solid, or of two sides of a plane figure.

Cross-hatched In mechanical drawing, a half tinting placed upon parts cut by a cutting plane. In free-hand drawing, the use of lines crossing each other and producing light and shade effects.

Curvature. Variation from straightness.

Curve A line of which no part is straight.

REVERSED. One whose curvature is first in one direction and then in the opposite direction

SPIRAL A plane curve which winds about and recedes, according to some law, from its point of beginning, which is called its *centre*.

Cylinder A solid bounded by a curved surface and by two opposite faces called bases , the bases may be ellipses, circles, or other curved figures, and name the cylinder Thus a circular cylinder (the ordinary form) is one whose bases are circles.

A RIGHT CIRCULAR CYLINDER is generated by the revolution of a rectangle about one side as an axis The side about which the rectangle revolves is called the *height* of the cylinder, also its *axis* The side opposite the axis describes the curved surface of the cylinder, and in any of its positions is called an *element* of the surface

Cylindrical. Having the general form of a cylinder

Degree The 360th part of a circumference of a circle.

Describe To make or draw a curved line.

Design Any arrangement or combination to produce desired results in industry or art

Develop To unroll or lay out upon one plane the surface of an object

Diagonal   A straight line in any polygon which connects vertices not adjacent.

In regular polygons, diagonals are called *long* when they pass through the centre, as *c d*, and *short* when they extend between parallel sides, as *a b*

Diameter.   See Circle   In a regular polygon with an even number of sides a line joining the centres of two opposite sides is often called a diameter, as *e t*.

Edge.   The intersection of any two surfaces   The boundary line. Edges are straight or curved, and are represented by lines

Elevation   A drawing made on a vertical plane by means of projecting lines perpendicular to the plane from the points of the object   The terms elevation, vertical projection, and front view all have the same meaning.

Ellipse   A plane figure bounded by a line such that the sum of the distances of any point in it, as *c*, from two given points *e* and *f*, called *foci*, is equal to a given line, as *a b*   The point midway between the foci is called the *centre*

> The Transverse Axis of an ellipse is the longest diameter that can be drawn in it, as *a b*.   It is also called the *major* or *long* axis
>
> The Conjugate Axis is the shortest diameter which can be drawn, as *c d*. It is also called the *minor* or *short* axis. The foci, *e* and *f*, are two points in the long diameter whose distance from *c* or *d* is equal to one-half *a b*

Face.   One of the plane surfaces of a solid   It may be bounded by straight or curved edges.

Finishing   Completing a drawing, whose lines have been determined, by erasing unnecessary lines and strengthening and accenting where this is required.

Foreshortening.   Apparent decrease in length, due to a position oblique to the visual rays.

Free-hand.   Executed by the hand, without the aid of instruments.

Frustum   See Cone and Pyramid.

Generated.   Produced by.

Geometric.   According to geometry   .

Half-tint   The shading produced by means of parallel equidistant lines

Hemisphere   Half a sphere, obtained by bisecting a sphere by a plane.

Horizontal.   Parallel to the surface of smooth water.

In drawings, a line parallel to the top and bottom of the sheet is called horizontal

Inscribe.   A polygon is said to be inscribed in a circle when all its vertices are in the circumference of the circle ; and a circle is said to be inscribed in a polygon when the circumference of the circle is touched by each side of the polygon.

Instrumental   By the use of instruments.

Lateral Surface   The surface of a solid excluding the base or bases

Line.   A line has length only.   In a drawing its representation has width but is called a line.

STRAIGHT.   One which has the same direction through-out its entire length.

CURVED.   One no part of which is straight.

BROKEN.   One composed of different successive straight lines.

MIXED   One composed of straight and curved lines

CENTRE.   A line used to indicate the centre of an object.

CONSTRUCTION.   A working line used to obtain required lines.

DOTTED.   A line composed of short dashes

DASH.   A line composed of long dashes.

DOT and DASH.   A line composed of dots and dashes alternating.

DIMENSION.   A line upon which a dimension is placed.

FULL.   An unbroken line, usually represent-ing a visible edge.

SHADOW.   A line about twice as wide as the ordinary full line.

A straight line is often called simply a line, and a curved line, a curve.

Longitudinal.   In the direction of the length of an object.

Model.   A form used for study.

Oblique.   Neither horizontal nor vertical.

Oblong.   A rectangle with unequal sides.

Oval   A plane figure resembling the longitudinal section of an egg ; or elliptical in shape.

Overall.   The entire length.

Ovoid.   An egg-shaped solid.

Parallel.   Having the same direction and everywhere equally distant.

Parallelogram.   See Quadrilateral.

Pattern.   That which is used as a guide or copy in making anything.

> FLAT.   One made of paper or other thin material.

> SOLID.   One which reproduces the form and size of the object to be made.

Perimeter.   The boundary of a closed plane figure.

Perpendicular.   At an angle of 90°.

Perspective.   The art of making upon a plane, called the *picture plane,* such a representation of objects that the lines of the drawing appear to coincide with those of the object, when the eye is at one fixed point called the *station point.*

> DIAGRAM.   An exact perspective drawing obtained scientifically by perspective methods.   It is often very false pictorially when not seen from the station point.

> PARALLEL   Diagram perspective which represents a cubical form by the use of one vanishing point, and represents by its real shape any face parallel to the picture plane.

> ANGULAR.   Diagram perspective in which two sets of horizontal edges of a cubical form are at angles to the picture plane, and the object is thus represented by the use of two vanishing points.

> OBLIQUE.   Diagram perspective in which, none of the edges of a cubical form being parallel to the picture plane, it is represented by the use of three vanishing points.

FREE-HAND or MODEL DRAWING. A drawing which, without confining the eye to the station point, represents as far as possible the actual appearance of objects. It is made free-hand, and is for most purposes more satisfactory than an exact diagram perspective.

Plan. Plan, horizontal projection, and top view have the same meaning, and designate the representation of an object made on a horizontal plane by means of vertical projecting lines. In architecture it means a horizontal section.

Plane Figure. A part of a plane surface bounded by lines.

A plane figure is called *rectilinear* if bounded by straight lines, *curvilinear* if bounded by curved lines, and *mixtilinear* if bounded by both straight and curved lines.

Similar figures are those that have the same shape.

Plinth. A cylinder or prism, whose axis is its least dimension. It is *circular*, *triangular*, *square*, etc., according as it has circles, triangles, squares, etc., for bases.

Polygon. A plane figure bounded by straight lines.

An EQUILATERAL POLYGON is one whose sides are all equal.

An EQUIANGULAR POLYGON is one whose angles are all equal.

A REGULAR POLYGON is one which is equilateral and equiangular.

PARALLEL POLYGONS are those whose sides are respectively parallel.

1.   2.   3.   4.   5.

TRIANGLE. A polygon having three sides (1).

QUADRILATERAL. A polygon having four sides (2).

PENTAGON. A polygon having five sides (3).

HEXAGON. A polygon having six sides (4).

HEPTAGON. A polygon having seven sides (5).

6.   7.   8.   9.   10.

The AXIS of a pyramid is a straight line connecting the vertex and the centre of the base.

The ALTITUDE of a pyramid is the perpendicular distance from the vertex to the base.

Quadrant. See Circle.

Quadrilateral. A plane figure bounded by four straight lines. These lines are the *sides*. The angles formed by the lines are the *angles*, and the vertices of these angles are the *vertices* of the quadrilateral.

A PARALLELOGRAM is a quadrilateral which has its opposite sides parallel.

A TRAPEZIUM is a quadrilateral which has no two sides parallel.

A TRAPEZOID is a quadrilateral which has two sides, and only two sides, parallel.

A RECTANGLE is a quadrilateral whose angles are right angles.

A SQUARE is a rectangle whose sides are equal.

A RHOMBOID is a parallelogram whose angles are oblique angles.

A RHOMBUS is a rhomboid whose sides are equal.

The side upon which a parallelogram stands and the opposite side are called respectively its lower and upper bases.

Quadrisect. To divide into four equal parts.

Quatrefoil. A figure composed of four leaf-like parts.

Section. A projection upon a plane parallel to a cutting plane which intersects any object. The section generally represents the part behind the cutting plane, and represents the cut surfaces by cross-hatching.

Sectional. Showing the section made by a plane.

Sector and **Segment.** See Circle.

Shadow. Shade and shadow have about the same meaning, as generally used ; but it will be well to designate by shadow those parts of an object which are turned away from the direct rays of light, while those surfaces which receive less direct rays and are intermediate in value between the light and the shadow are called shade surfaces

CAST   The shadow projected on any body or surface by some other body.

Similar Figures are those which have the same shape

Solid. A solid has three dimensions, length, breadth, and thickness It may be bounded by plane surfaces, by curved surfaces, or by both plane and curved surfaces As commonly understood, a solid is a limited portion of space filled with matter, but geometry does not consider the matter and deals simply with the shapes and sizes of solids

Sphere. A solid bounded by a curved surface every point of which is equally distant from a point within called the centre.

A sphere may be generated by the revolution of a circle about a diameter as an axis

Spheroid (Ellipsoid) A solid generated by the revolution of an ellipse about either diameter. When revolved about the long diameter, the spheroid is called *prolate*, or the long spheroid; when about the short diameter, it is called *oblate*, or the flat spheroid The earth is an oblate spheroid

Spiral. See Curve.

Surface The boundary of a solid. It has but two dimensions, length and breadth

Surfaces are plane or curved

A PLANE SURFACE is one upon which a straight line can be drawn in any direction.

A CURVED SURFACE is one no part of which is plane.

The surface of the sphere is curved in every direction, while the curved surfaces of the cylinder and cone are straight in one direction

The surface of a solid is no part of the solid, but is simply the boundary of the solid It has two dimensions only, and any number of surfaces put together will give no thickness

Symmetry *Design* A proper adjustment or adaptation of parts to one another and to the whole

BILATERAL.   Having two parts in exact reverse of each other

Symmetry. *Geometry.* If a solid can be divided by a plane into two parts such that every straight line, perpendicular to the plane and extending from the plane in each direction to the surface of the solid, is bisected by the plane, the solid is called a *symmetrical* solid, and the plane is called a *plane of symmetry.* If two planes of symmetry can be drawn in a solid, their intersection is called an *axis of symmetry.* See Axis of Symmetry.

The line about which a plane figure revolves when it generates a solid of revolution is an axis of symmetry for the solid; it is also called the *axis of revolution.*

Tangent. A straight line and a curved line, or two curved lines, are tangent when they have one point common and cannot intersect; lines or surfaces are tangent to curved surfaces when they have one point or one line common and cannot intersect.

Trefoil. A figure composed of three leaf-like parts.

Triangle. A plane figure bounded by three straight lines. These lines are called the *sides.* The angles that they form are called the *angles* of the triangle, and the vertices of these angles, the *vertices* of the triangle.

Triangles are named by their sides and angles.

A SCALENE TRIANGLE is one in which no two sides are equal.

An ISOSCELES TRIANGLE is one in which two sides are equal.

An EQUILATERAL TRIANGLE is one in which the three sides are equal.

A RIGHT TRIANGLE is one in which one of the angles is a right angle.

An OBTUSE TRIANGLE is one in which one of the angles is obtuse.

An ACUTE TRIANGLE is one in which all the angles are acute.

The HYPOTENUSE is the side of a right triangle opposite the right angle. The other sides are called the *legs.*

An EQUIANGULAR TRIANGLE is one in which the three angles are equal   The value of each angle is 60°.

The BASE is the side on which the triangle is supposed to stand. In an isosceles triangle, the equal sides are called the *legs*, the other side the *base;* in other triangles any one of the sides may be called the base

The ALTITUDE is the perpendicular distance from the vertex to the base.   Except in the isosceles triangle, there are three altitudes.

The vertex of the angle opposite the base is often called the *vertex* of the triangle.

Trisect.   To divide into three equal parts.

Truncated.   A truncated solid is the part of a solid included between the base and a plane cutting the solid oblique to the base.

Type Form   A perfect geometrical plane figure or solid.

Vertical.   Upright or perpendicular to a horizontal plane or line.

Vertical and perpendicular are not synonymous terms.

Vertex.   See Angle, Quadrilateral, Triangle.   The vertex of a solid is the point in which its axis intersects the lateral surface.

View.   See Elevation.   Views are called front, top, right or left side, back, or bottom, according as they are made on the different planes of projection.   They are also sometimes named according to the part of the object shown, as edge view, end view, or face view.

Working Drawing   One which gives all the information necessary to enable the workman to construct the object.

Working Lines.   See Lines.